T0269053

Ecological and environmental education in schools and institutes of further and higher education has gained increased importance in recent years, both as an area of study in its own right, and as a component of other disciplines. There is now a requirement in many countries to include the environment in both formal and informal curricula.

This volume presents a long overdue account of the status, progress and underlying concepts of ecological education. It explores areas of recent development and debate in ecological and environmental education, describes the evolution and development of environmental education in different countries, and examines the importance and provision for fieldwork. Case examples illustrate how ecological studies are undertaken in several culturally different settings.

This book will interest teachers and research workers in ecology, environmental science and education.

# ECOLOGY IN EDUCATION

# ECOLOGY IN EDUCATION

*Edited by*

MONICA HALE

CAMBRIDGE
UNIVERSITY PRESS

CAMBRIDGE UNIVERSITY PRESS
Cambridge, New York, Melbourne, Madrid, Cape Town, Singapore, São Paulo, Delhi

Cambridge University Press
The Edinburgh Building, Cambridge CB2 8RU, UK

Published in the United States of America by Cambridge University Press, New York

www.cambridge.org
Information on this title: www.cambridge.org/9780521556699

First published 1993
First paperback edition 1995

*A catalogue record for this publication is available from the British Library*

*Library of Congress Cataloguing in Publication data*

Ecology in education / edited by Monica Hale.
    p.    cm.
Based on a symposium held in Yokohama, Japan, in Aug. 1990.
Includes bibliographical references.
ISBN 0-521-43346-0 (hc)
1. Ecology – Study and teaching – Congresses.   2. Environmental
education – Congresses.   I. Hale, Monica.
QH541.2.E25   1993
574.5′071 – dc20   93-26183   CIP

ISBN 978-0-521-43346-4 hardback
ISBN 978-0-521-55669-9 paperback

Transferred to digital printing 2009

# Contents

# Foreword: General understanding and role of ecology in education

FRANK B. GOLLEY

*Research Professor of Ecology, University of Georgia, and Past President, International Association for Ecology, INTECOL*

Beginning in the 1960s the word 'ecology' began to take on a life of its own. I recall, as a director of an Institute of Ecology, drafting a letter to editors of newspapers and magazines who had misused the word. The letter defined ecology properly and pointed to their mis-usage. It ended with an offer to discuss the appropriate language of the environmental movement over the phone or to send them materials on scientific ecology. There was no response to the offer! We still hear high placed administrators of the environment telling us that he or she has implemented policies to save 'the ecology'.

Confusion about language characterises our time. Presidents call war peace, we are told that tyranny is democracy, capitalism is termed socialism and on and on. We live in a time when confusion over moral purpose, over the role of government and the rights and responsibilities of citizenship seem ubiquitous. We have had more than 30 years of intense discussion and research on the environmental crisis, yet the crisis becomes worse daily. Savants tell us we live in a time of transition, a time when the modern world in which we were born and have lived is being transformed into something different. We all hope, of course, that this new state will incorporate the positive qualities of the familiar modern world and eliminate the negatives. Certainly, for a positive path of transformation in this post-modern period we will depend upon the educational system. Education must provide the young with the tools they need to become effective members of their society and to provide adults with new information so that they can adapt to changing circumstances.

In other words, a key, possibly the most important key, to successful transformation is education. Ecological or environmental education is a fundamental part of the total educational package, especially in the primary and secondary grades. If a student does not understand the concept of environment and their personal relations to the environment, they are illiterate. Every judgment we make, at every moment of time, has an environmental component. It is essential that we teach the young that they are responsible for their environmental actions and that irresponsibility feeds back to them in terms of ill health, lack of performance, misery and ultimate poverty. Social action is linked tightly to environmental action.

However, ecological education must also be continued at the higher levels of education as well. Here is where the specialists who will manage the environment and study it are trained. Ecological systems are so complex that study of them is continually required. We never stop needing an understanding of how our environment changes and affects us. Further, those students concerned with medicine, engineering, business and other technical and applied subjects need to understand how the environment interacts with their professional interests. Every field has an environmental component. At my university, The University of Georgia, Charles Knapp, our President, declared in his State of the University Address that he wanted us to develop programs to ensure that all graduates of the university were environmentally literate. The faculty has been studying how to implement this goal for more than a year. It is neither simple nor easy to create the interest and will to consider the environmental interactions within the disciplines. Yet, our committees have been enthusiastic that it can be done. They have recommended that faculties study and search for the environmental connections in each department and school. They have avoided the dogmatic environmental position and have chosen diversity and flexibility. I think that this is the best choice. Who knows what we will be required to teach about the environment in the next decade and the one after that? Who would have predicted the rapid deterioration of the ozone layer or the synergistic response of pollutants? Our environment and our society is continually dynamic.

But the ecological and environmental education specialists have an even broader agenda. They include within their interests adult education too. We have long passed the point where a few years of schooling and then many years of job experience is sufficient to equip a person to cope with a world of industrial, social and technical change. People need continual education. This need is especially true in the environmental area because the environment is so rapidly changing and dynamic and our understanding of the environment advances rapidly. We can think of our relationship to the environment through an analogy with the oriental martial arts. For successful adaptation, we act with disturbance or change and seek to direct and guide it into the paths we choose. We do not attack or seek to directly stop change. This kind of environmental management requires well educated practitioners.

We need an integrated ecological education starting with the primary grades and extending continuously into adult life. The child learns why we are concerned about the environment and the basic principles of ecology. The college student learns how to study the environment and how to apply advanced knowledge to other disciplines. The employed adult learns the latest environmental knowledge in order to apply it in practice. I wonder if any society has such a broad program in place?

Finally, let me return to the theme of the Congress of Ecology, which was the stimulus for writing the papers that make up the chapters of this book.

Ecology is a science that developed in Europe and North America. It has become an important part of science in universities that teach a western orientated, modern education curriculum. However, the human-environment relationship is universal. It develops through the interaction of humans and their environments and human adaptation to the environmental circumstances of each place and time. While in one sense ecological science is universal, in another sense it is culturally relative to the place of its origin. Its principles are universal. The applications of those principles are local and specific. Therefore, there is a rich field of ecological application that underlies ecological education. The two must fit together otherwise the lessons may be inappropriate and possibly destructive. It was exciting for me to see this theme of cultural relativity worked out in the papers of the ecological education sessions at the Congress of Ecology.

Ecological education is a key part of the discipline. We hope that this present volume will continue the advance to create an environmentally literate society.

# Preface

The global environment has deteriorated to such an extent that the great life-supporting systems of the planet's biosphere are being threatened. Changes in the earth's climate, the accumulation of waste products, soil exhaustion, deforestation and the destruction of ecosystems, are already apparent.

The radical change in human attitudes forseen by the acceptance of the concept of sustainable development depends on a vast campaign of public education and re-education, a world-wide debate around these vital life and death issues to start now if sustainable human progress is to be achieved.
(*Gro Harlem Brundtland, former Prime Minister of Norway, Chairman of the World Commission on Environment and Development (1987)*)

The Fifth INTECOL Congress held in Yokohama, Japan, in August 1990 focussed on the 'Development of ecological perspectives for the 21st century'. This was the first international congress on ecology to be held in Asia. Almost 3000 ecologists and environmental professionals from in excess of 80 countries participated in over 100 symposia and plenary sessions.

The chapters in this book are based on the congress symposium 'The general understanding and role of ecology in education', which brought together ideas, achievements and current developments in ecological and environmental education from many parts of the world. The symposium's aim was to stimulate interest and commitment from academic ecologists to broaden the base of the dissemination and application of ecological science on a wider educational basis.

There has not been a significant review of international ecology education since the symposium, 'The teaching of ecology' in 1966 organised by the British Ecological Society (BES). In Lambert's introduction to the symposium volume she stated that 'Ecology is rapidly gaining ground as an integral part of modern education', while noting that ecology is 'a subject which seems to have given rise to more controversy as to its place in our education system than any other branch of biology' (Lambert, 1967). A quarter of a century later, this is not now the case as ecology is taking on a more integrated role and is included in a broad range of subjects: not just confined to biology but also finding a place in geography, environmental science, the broad-based disciplines and subjects and in environmental education generally.

Ecology is now viewed as a basic component of education to be included in the curriculum for all, not just those studying biology or going on to higher and further education. There has been a rapid rise in the public's general interest in the 'natural world' demonstrated by increasing media attention, numerous television programmes and books on ecological topics and issues. A greater appreciation of the world around us has taken root and is a trend that is growing internationally.

Professor Frank Golley (President of INTECOL 1986–90) in his opening address to the 1990 Congress emphasised that ecological education, as one of the most pressing needs for further development in the 1990s, will be vital if the twenty-first century is not to suffer ecological and social disaster on a global scale. That *all* people of the world should be ecologically literate is an aim that must be strived for.

It is therefore timely to compare and consider developments in other parts of the world and to review the different approaches adopted.

The papers reproduced in this volume provide an account of how ecology in education has developed since the BES Symposium on ecology education in 1966. They describe recent advances and approaches adopted in different educational settings in a range of environments, for various age groups, education systems, curricula and syllabuses in different parts of the world. The contributions include considerations of the basic problems of providing opportunities for ecological education for all young people and adults through the various levels and stages of education.

The chapters reflect the wide range of work currently being undertaken in ecological education and changes in environmental philosophy. This is borne out in particular by Robottom, who argues that the scientific, technocratic emphasis of ecology and the solving of environmental problems has diminished our capacity to deal effectively with them. He hypothesises that failing to understand environmental problems as essentially political issues has resulted in a failure to provide an appropriate form of environmental education in schools.

Ecology education in both formal and informal education at different levels (school to adult education) is explored in terms of its place in the curriculum, its content and the ways in which it may be developed. Examples from developed and developing countries are reviewed in the chapters by Acar, Thomas, and Hale and Hardie.

Berkowitz examines the status of ecology education in the curriculum in the United States and describes some of the inherent difficulties in organising ecological study programmes at school level. He then goes on to show how these difficulties are being overcome with the support of outside expertise, such as involving ecologists and specialist organisations in specific programmes of ecology education.

Field-work has traditionally been an integral part of ecological and environmental studies. This can act to discourage some from undertaking practical work as the outside environment often presents a bewildering array of phenomena and modes of study to both teachers and students alike. Tilling reviews the problems and constraints encountered in teaching field ecology at school level and examines the relevance of practical experience to higher education and research in ecology and other areas of study. The differences in field-work provision between developing and developed nations are considered by Leal Filho. From a survey of schools in Britain and Brazil, Leal Filho shows that the disparities that exist appear not only as a result of differences in funding support but traditional methodology and historic factors also play a major role in the level of provision of field experience.

The benefits of practical experience in the environment are demonstrated in a comparative study carried out by Harvey, who assesses the effects of contact with vegetation on how children view their immediate surroundings. The benefits arising from this experience underline the need to ensure children are given every opportunity to study and play in the outdoor environment. This theme is taken up by Adams in reviewing the provision of school sites for experiential learning. As well as the school site itself, the local environment can provide a wealth of opportunities to undertake ecological and environmental study. As Hale shows, the importance of the urban environment has only comparatively recently been recognised as providing a stimulus for ecological studies. As much of the world's population is concentrated in urban areas and, by implication, is potentially out-of-touch with the 'natural world', the development of field-work in the built environment is of particular importance.

Ecological education has taken a number of forms and has developed at varying rates in different countries. Cultural factors have been shown to influence the pace of development and the nature of ecological education. Soerjani and Yoshida review the development of ecological education in two Asian countries, Indonesia and Japan respectively.

The promotion of projects devised and coordinated by organisations and individuals outside the formal education sector has been a common feature of ecological education. These projects are often centred around a theme or issue and impart expertise and techniques to pupils or communities. Two examples of this strategy are described in relation to a comparative study of acid rain in Norway and Britain (by Shirley) and conservation education in West Africa (by Smith).

This book is not a manual on how to teach ecology, nor does it present unified solutions to the many ecological and environmental problems of the day. Indeed, the opinions of the various contributors frequently differ: the mode of thought or action adopted by one is sometimes a direct antithesis to that considered or found to be successful by another. Essentially, the chapters review and debate aspects of ecological education and discuss issues where no clear agreement has been reached by practitioners.

The fact that there are many views and approaches to ecology education is itself an indication of the vitality of ecology today. As has been observed, 'Ecology is still essentially a "young" subject in terms of established disciplines'. Indeed, the world's oldest ecological society (the British Ecological Society) celebrated its 75 years Jubilee only in 1988.

All that this volume can do is to offer suggestions for consideration in the light of individual circumstances and experience; it will have served its purpose if it leads to a recognition of the increasing part that ecology must play in the general field of education.

The symposium papers are an important bridge between the first 'Teaching of Ecology' conference organised by the BES in 1966 and future work that will be vital to the well-being of every nation and to the world as a whole. Only through ecological knowledge and understanding and the engendering of skills of interpretation and good environmental housekeeping, can people develop a sense of concern for what is happening on a local and global scale and be encouraged to take appropriate action.

Monica Hale

## References

Lambert (1967) *The Teaching of Ecology*, Oxford: Blackwell.
World Commission on Environment and Development (1987) *Our Common Future* (The Brundtland Report) Oxford.

## Acknowledgements

The organiser and contributors to the symposium 'General understanding and role of ecology in education' of the Fifth INTECOL Congress are indebted to the INTECOL Congress Organising Committee and the Japanese Local Program Committee for hosting the congress and for the excellent arrangements made to ensure the congress achieved maximum success.

Thanks in particular go to Professors Yuiti Ono, Paul Maycock and Frank Golley for their hard work prior to and during the congress and to Dr T. Yoneda who co-chaired the Ecology in Education Symposium.

The generous financial support of PowerGen, The British Ecological Society, The Department of Education and Science, The British Council and the INTECOL Congress Fund is gratefully acknowledged in making the symposium possible.

# 1

# The role of ecology in education: an Australian perspective

IAN ROBOTTOM

*Associate Professor in Education, Faculty of Education, Deakin University, Geelong, Victoria 3217, Australia*

## Introduction

This chapter will begin with a reflection on the field of environmental education in the last decade or so. It will be argued that the field has been dominated by a 'technocratic rationality' – a technocratic curriculum with an emphasis on the transmission (as propositional knowledge) of objectivist conceptions of 'ecology' and other scientific concepts that tend to limit our understanding of environmental issues; and that technocratic approaches to teacher education in environmental education (with an emphasis on top-down or center–periphery processes of professional development) tend to limit the opportunities for teachers to explore alternative forms of curriculum in environmental education.

## Technocratic rationality in environmental education

One of the myths in environmental education is that its origins lie in the field of science education or the practice of nature study – that environmental education is in some sense a step-child of science education. The records of the UNESCO–UNEP International Programme in Environmental Education show that environmental education as we know it today originated jointly in the concerns of Third World countries about the extent to which their countries were being degraded environmentally by the activities of developed (overdeveloped?) countries, and in the concerns of fairly ad hoc community groups about the visibly deteriorating environment. These origins were essentially political in nature, involving the often competing vested interests of individuals, groups and nations.

Yet ironically, as the environmental education movement became established and began to express itself in school curricula, there was a marked tendency for environmental education to be co-opted by science education, with the result that much of the former's political edge was lost. Environmental problems became interpreted as mainly technical problems susceptible to technical solutions of the kind that science is competent in supplying. Environmental education came to be dominated by the natural science approach (Schleicher, 1989). A heavy emphasis emerged on the formal provision of systematic knowledge drawn from the traditional science disciplines of geography, geology and biology (especially the treatment of basic ecological principles).

It seemed to be assumed that the acquisition by students of an ecological perspective and an appreciation of the ecosystem concept would forestall environmentally damaging actions on their part, and develop in them an informed concern for the environment.

The irony in this trend is that to the very extent that the problems and solutions came to be seen from a scientific perspective (that is, to the extent that environmental education came to be seen as an applied science), our rationality came to be of the *technocratic* kind. Technocratic rationality tends to be marked by a dominant and almost blind faith in the capacities and qualities of science (prime among these being claims about 'objectivity', 'rationality' and 'truth') to deal effectively and efficiently ('positively') with a range of problems that beset us. As this dominant technocratic rationality subsumed the fledgeling environmental education movement, there was a diminishment of the important capacity to see environmental problems as essentially political issues to do with contests between differing vested interests. In the author's view, failing to understand environmental problems in these terms in turn diminishes our capacities to deal effectively with them, and to provide an appropriate form of environmental education in schools. As early as 1977, an influential UNESCO publication recognised and described the danger of this 'technocratic rationality' in environmental education:

At any one time, the educational system – whether based on religious dogmas and practices or on rational thought – has tried to divulge, sustain, and perpetuate sets of social values. The process has occurred at some times openly, at other times through devious channels. If you consider the world today and examine the diverse educational systems, you can clearly identify competing ideologies; those which are attempting to hold on to recognised and almost undisputed values, and those which have launched a major strategy for conquering the world and men's minds.

In other terms, behind any educational process lies a philosophy, a moral philosophy, for the people who exert power and are in charge of educational institutions share certain values, which they wish to disseminate in order to ensure the prolongation, if not the indefinite survival, of the system they are devoted to. *(Buzzati-Traverso, 1977, p. 14)*

One of the outcomes of a technocratic rationality in education in general is an emphasis on the didactic teaching of pre-ordinate knowledge – knowledge that is systematically selected and organised before the classroom activities through which it is 'transmitted' to students. In environmental education, notions like 'the ecosystem concept' and 'basic ecological principles' form a conspicuous part of this pre-ordinate knowledge. In this approach, 'ecology' is often treated as a means of perceiving the environment 'as it is', as it 'really exists out there' in a purportedly objective sense, in a way that is disjoined from personal, political and social values.

'Ecological principles' are perceived as the framework that we need in order to 'see' the environment as it really is – as if the environment is a fixed, concrete entity awaiting discovery by people equipped with the right tools. That is, technocratic rationality is expressed not only in the pedagogy of environmental education (transmission of propositional knowledge) but also in the objectivist images of 'environment' and 'ecology' that tend to be promulgated. This is in contrast with more subjectivist post-empiricist views (for example, in some recent formulations of 'science–technology–society' (STS) – see Aikenhead, 1988) in which 'ecology' may be seen as an essentially human figment, concerned with the development and application of a *socially constructed framework* in a process of *interpretation of an experienced environment*. In the words of Giovanna Di Chiro:

The environment is what surrounds us, materially and socially. We define it as such by use of our own individual and culturally imposed interpretive categories, and it exists as the environment at the moment we name it and imbue it with meaning. Therefore, the environment is not something that has a reality totally outside or separate from ourselves and our social milieux. Rather it should be understood as the conceptual interactions between our physical surroundings and the social, political and economic forces that organise us in the context of these surroundings. And if we view the environment as a social construct then we accept that certain qualities of it can be transformed according to whichever social relationships are in operation.

If we view the environment as a social construct, we can also view the 'environmental problem' very differently ... Environmental problems are ... social problems, caused by societal practices and structures, and only viewed or socially constructed as problems because of their effects on human individuals and groups (of course other living things and systems are also affected). *(Di Chiro, 1987, p. 25)*

To the extent that environmental education programs are based on the development of an objectivist understanding of 'ecology' – as an independently existing, fixed and 'real' framework – there is a developed blindness to the fundamentally political character of environmental problems. Put simply, if environmental problems are described only in the scientific terms of objective relationships between physically existing components, then important factors like vested human and state-related interests are overlooked:

The social component is all the more important because man tries to adjust nature in his interest and changes the environment according to value preferences, and because no lasting 'meaning' can be attributed to nature.
*(Schleicher, 1989, p. 62)*

Awareness of environmental problems is social awareness rather than ecological awareness. Such problems will be solved through collective action aimed at eradicating the social and economic causes of the degradation of the human environment. The political aspects of this search for solutions may give rise to conflicts of various kinds. One such conflict, and not the least, is the collision between the educational system and the private interests which operate in alliance with the powers of the State.
*(Vidart, 1989)*

Another outcome of technocratic rationality in environmental education is a belief in the authority of scientific knowledge, which expresses itself in various divisions of labour: for example, divisions between those who would produce knowledge (this tends to be the scientific academy) and those who ought to use or implement knowledge (practitioners of various kinds). In education, a technocratic rationality supports the division of *theorising, research and development* on the one hand (this being seen as the proper domain of the 'academy'), and *teaching practice* as technical implementation on the other (this being seen as the proper domain of schools). This division of labour is clearly seen in the 'research, development, diffusion, adoption' (RDDA) model of professional development and curriculum development in environmental education (Robottom, 1987a).

These two outcomes tend to interact. The technocratic interest that justifies and preserves a division of labour between the science (or science education) academy and teachers creates the conditions for the academy to enact a role of legitimating pre-ordinate, objectivist ecological knowledge as proper curriculum content. This can be seen in the move to a National Curriculum in the United Kingdom and, more recently, in Australia. However, as we shall see, there are recent developments that challenge the dominant technocratic rationality in environmental education.

## Recent international environmental education-related developments

### (i) 'Science–technology–society'

There is increasing worldwide interest (see for example: Aikenhead, 1988; Iozzi, 1987; McFadden *et al.*, 1989) in a relatively new educational initiative, 'Science–technology–society' (STS). This approach has come to mean a way of teaching scientific content and skills in a meaningful context of technology and society. Teaching about relationships between science, technology and society is a relatively recent departure from former approaches in which science education in schools was seen in the vocational sense of preparation of students for technological competence in the society of the day.

Some narrow interpretations of STS have simply sought to teach those aspects of conventional science content that seem to have some application in alleviating certain pressing social problems. These interpretations suggest a one-way, instrumental relationship between science and society (science 'contributes' solutions to society), and tend to reinforce the traditional mainstay of science – its claim to objectivity, rationality and truth. But perhaps the real potential in STS is to teach science in a way that challenges rather than reinforces that traditional mainstay – to create the conditions for students to understand the social structure of science itself (Kuhn, 1962). There is the opportunity for activities that demonstrate how science (its research topics; what counts as appropriate research questions; what counts as acceptable methodologies and outcomes) is as much influenced by the society of the day as science itself shapes society by the creation of knowledge and the provision of solutions to problems. The STS initiative supplies opportunities for the truth claims of science to be presented as requiring appraisal in terms of their historical and cultural context.

The relevance for environmental education of these STS developments is that they offer a new way of perceiving science and its relationship with society – a way that allows us to shed some of the shackles of technocratic rationality. If science is perceived as socially constructed, then its special claims about the objectivity, rationality and truth of its knowledge (and the scientific processes that yield that knowledge) can be challenged. This may be a pre-requisite condition for recognising that the concepts of

'ecology' and 'environment', themselves part of the language of the traditional science discipline, are also social constructions and perhaps can be
taught in environmental education in other than didactic, propositional
fashion (see also Schleicher, 1989).

## (ii) Practitioner research in environmental education

Kathleen Kelley of the Centre for Educational Research and Development at the Organisation for Economic Co-operation and Development
(OECD) in Paris coordinates a current major OECD-funded project
in Western Europe, the 'Environment and School Initiatives' project
(see Posch, 1990). This project has two important dimensions: a substan-
tive environmental education emphasis on participatory, action-based
environmental enquiries; and a 'procedural', professional development
emphasis on systematic reflection on action by participating teachers
(action research). What is distinctive about this project is that it argues
for and enacts a higher professional role for practitioners – rather than
confining teachers to the role of technical implementers of the curricula
designed by others (as is the case in the RDDA approach), the project
encourages teachers to participate in research of their own, conducted in
their own classrooms, and addressing environmental education issues of
interest and concern to themselves.

   Robottom and Muhlebach (1989), Greenall Gough and Robottom
(1993), and Robottom (1990) describe a project in Australia in which
students and teachers in seven isolated coastal schools participated in
three overlapping activities: (i) enquiries into controversial issues concerning the quality of local freshwater and marine environments; (ii) participation in an international computer conference on the subject of water
quality; and (iii) engagement in participatory educational research into
the pedagogical and curriculum issues that arise as attempts are made to
demonstrate the first two dimensions of the project, in a similar fashion
to the action research conducted by the teachers in Kelley's study. One of
the understandings emerging from this project is that, in environmental
education, there may be a surprisingly minor role for taught ecological
principles in contrast with opportunities for engaging social and political
influences. Another understanding is the importance of the 'working
knowledge' that emerges from the critical, community-based enquiries of
students as they investigate local, controversial environmental issues.
Action-based, community-embedded forms of enquiry yield knowledge
that is transactional rather than transmissional, generative/emergent

rather than pre-ordinate, opportunistic rather than systematic, and idiosyncratic rather than generalisable. This emergent working knowledge needs to be recognised as proper curriculum 'content' with at least equal status to pre-ordinate, systematic ecological principles. The Australian project demonstrates that this 'working knowledge' may take several forms – for example: computer conference dialogue in which students at the same or different schools articulate methodological problems as well as outcomes of environmental investigations; newspaper cuttings describing the influence in the community of the environmental enquiries conducted by students and teachers; and students' and teachers' contributions to workshops and conferences.

Emerging from these and other international perspectives on environmental education (see also Hale, 1990; Hart, 1990) is a recurring concern that in environmental education we need to enact alternatives to the dominant technocratic rationality of traditional approaches to curriculum development and professional development. Two ways in which these alternatives may be expressed are:

*Redefining curriculum content: recognising that in addition to the pre-ordinate, systematically organised and presented propositional knowledge (such as 'basic ecological principles') drawn from the traditional fields of knowledge (such as biology), there is value in the propositional 'working knowledge' that emerges from the socially critical enquiries of students and teachers as they conduct authentic research into local, controversial environmental issues in their communities. Of course, because such knowledge is community-based, it is idiosyncratic and does not fit well with notions of universal, generalisable curriculum content (Greenall Gough and Robottom, 1993).

*Role of the academy: recognising that the academy (researchers and teacher educators at colleges and universities; upper and medium level consultants in departments of education) may need to redefine their role. Their role may need to change from one of agency in the technocratic (RDDA) model of professional and curriculum development in which the problem of educational change is perceived as one requiring transmission of centrally determined 'solutions' in the form of curriculum materials to the 'periphery' (teachers in schools), to a role of creating the supporting conditions for teachers and others in their respective communities to carry out their own critical reflective enquiries into *their* theories, practices and educational predicaments, and the relationships between these (see Posch, 1990; Robottom, 1987b).

## Conclusion

Politicised expressions of environmental education are hampered by the co-optation of the field by the technocratic rationality of empirical/analytic science. This is especially evident in the heavy dependence in environmental education curricula on the systematic treatment of 'basic ecological principles' as a key body of knowledge, and in professional and curriculum development in environmental education. Recent international developments in the field suggest that the relationship of some of the mainstays of curriculum (for example: the notion of universal, pre-ordinate curriculum content on ecology; the RDDA approach to educational change) with environmental education are in need of review.

## Note

A modified version of this chapter has been published in the Canadian Journal of Experiential Education 14(1): 20–26, 1991.

## References

Aikenhead, G. (1988). *Teaching Science through a Science-Technology-Society-Environment Approach: An Instructional Guide*. Regina, Saskatchewan: Saskatchewan Instructional Development and Research Unit.
Buzzati-Traverso, A. (1977). Some thoughts on the philosophy of environmental education. In *Trends in Environmental Education*. ed. UNESCO, pp. 13–19. Paris: UNESCO.
Di Chiro, G. (1987). Environmental education and the question of gender: a feminist critique. In *Environmental Education: Practice and Possibility*, ed. I. Robottom. Geelong, Victoria: Deakin University Press.
Greenall Gough, A. and Robottom, I. (1993). Towards a socially critical environmental education: water quality studies in a coastal school. *Journal of Curriculum Studies*, **26**(2), (in press).
Hale, M. (1990). Recent developments in EE in Britain. *Australian Journal of Environmental Education*, **6**, 29–44.
Hart, P. (1990). Environmental education in Canada. *Australian Journal of Environmental Education*, **6**, 45–66.
Iozzi, L. (1987). *Science-Technology-Society: Preparing for Tomorrow's World*. New Brunswick, NJ: Cook College, Rutgers University.
Kuhn, T. (1962). *The Structure of Scientific Revolutions*. University of Chicago Press.
McFadden, C, *et al.* (1989). *Science Plus Technology and Society*. Toronto: HBJ-Holt.
Posch, P. (1990). Educational dimensions of environmental initiatives. *Australian Journal of Environmental Education*, **6**, 79–92.
Robottom, I. (1987a). Two paradigms of professional development in environmental education. *The Environmentalist*, **7**(4), 291–298.

Robottom, I. (1987b). Towards inquiry-based professional development in environmental education. In *Environmental Education: Practice and Possibility*, ed. I. Robottom, pp. 83–110. Geelong, Victoria: Deakin University Press.

Robottom, I. (1990). Reconstructing the curriculum for environmental responsibility. *New Education*, **XII**(1), 61–71.

Robottom, I and Muhlebach, R. (1989). Expanding the scientific community in schools: a computer conference in science education. *Australian Science Teachers Journal*, **35**(1) 39–47.

Schleicher, K. (1989). Beyond environmental education: the need for ecological awareness. *International Review of Education*, **35**(3), 257–281.

Vidart, D. (1989) Environmental education – theory and practice. *Prospects*, **VIII**(4) 466–479.

# 2

# Ecology and environmental education in schools in Britain

MONICA HALE

*Faculty of Human Sciences, London Guildhall University, Calcutta House, Old Castle Street, London E1 7NT, UK*

JACKIE HARDIE

*Deputy Head (Curriculum), The Latymer School, Haselbury Road, Edmonton, London N9 9TN, UK*

## Introduction

The past decade has seen a marked increase in public awareness of the natural world, mainly due to increased attention given by the media to natural history and related topics. Threats to the environment are becoming more visible as a consequence of this focus by the media.

The successful management of the environment in the future depends on the actions of government, industry, society and individuals. Sustainable, environmentally sound development will become a reality only if public awareness is coupled with the appropriate knowledge and skills and positive attitudes towards the environment.

Education has a key role to play in achieving a sustainable economy and society. This was emphasised by the Brundtland Commission in its report *Our Common Future* (World Commission on Environment and Development, 1987) in calling for a 'vast campaign of education, debate and public participation' to 'start now if sustainable human progress is to be achieved'. Present and future generations of school children are entitled to and must have access to a curriculum that addresses such matters.

Environmental education covers a broad spectrum and courses have been designed in the past whereby students receive their education, through, about, in and for the environment.

Ecological education has a narrower focus than environmental education and has been present in the curriculum of UK schools for some time. Its study was usually restricted to those following examination courses in biology or to those who were being taught by teachers with a particular interest in the environment. The Education Reform Act (1988), affecting state schools in England and Wales, ensures that all pupils of compulsory school age (5 to 16 years) will follow a balanced curriculum, and certain ecological and environmental principles are incorporated into the science and geography components. Thus, the opportunity to study ecology and environmental education is an entitlement for all.

## The nature of ecology

Recently, Malcolm Cherrett organised a survey of the British Ecological Society's* membership to determine the understanding of modern ecology. Respondents were asked to list the ten principal concepts that comprise modern ecology in order of importance (Table 2.1; Cherrett, 1989; Hale, 1991).

Of the 50 concepts identified by professional ecologists as integral to ecological science, few are currently included in the examination syllabuses at 16+ (General Certificate of Secondary Education – GCSE) for science and biology, or in the National Curriculum for science and geography. However, there are a substantial number of environmental topics included in the science and geography GCSE and in the National Curriculum which are not included in the ecologists' fundamental concept list as reported by Cherrett (op. cit.) (Table 2.2).

Ecologists do not agree on the most important ecological concepts; however, Cherrett's work illustrates the wide range of special interests of ecologists and, as a consequence, ecology, as a science does not appear to have a unifying philosophical or theoretical base, unlike physics or chemistry.

As Cherrett (op. cit.) believes, this presents a problem of communication – how ecological science should be explained to the wider, lay world, and, in relation to schools, how to target discussion which may influence the curriculum.

## Environmental education in the UK

Before the Education Reform Act (1988) teachers in the UK had considerable freedom related to the choice of content of courses. This resulted in variety of provision and many exciting developments, often funded and supported by local government. Conversely, some children were denied access to certain curriculum areas and ensuring continuity of experience and progression in learning was problematic. The Education

* The British Ecological Society, 26 Blades Court, Deodar Road, Putney, London, SW15 2NU, UK.

Table 2.1. *Principles of ecology ranked in order of importance and compared with the ecological content of the National Curriculum and GCSE syllabuses for science and geography.*

| BES rank | Principal concepts of ecology | GCSE | | Natl. Curr. | | |
|---|---|---|---|---|---|---|
| | | Sci. | Biol. | Sci. | Geog. | XCurr. |
| 1 | The ecosystem | 10 | 7 | | (AT3) | * |
| 2 | Succession | 1 | 3 | | | * |
| 3 | Energy flow | 9 | 6 | AT2 | | * |
| 4 | Conservation of resources | 7 | 6 | AT2 | AT5 | * |
| 5 | Competition | | 4 | | | |
| 6 | Niche | | | | | |
| 7 | Materials cycling | 6 | 6 | AT2 | | * |
| 8 | The community | | | | | |
| 9 | Life history strategies | 9 | 4 | AT2 | | |
| 10 | Ecosystem fragility | | | | | |
| 11 | Food webs | 10 | 6 | | | |
| 12 | Ecological adaptation | | | | | |
| 13 | Environmental heterogeneity | | | AT2 | | |
| 14 | Species diversity | 7 | 6 | AT2 | | * |
| 15 | Density-dependent regulation | | | | | |
| 16 | Limiting factors | 1 | 2 | | | |
| 17 | Carrying capacity | | 4 | | | * |
| 18 | Maximum sustainable yield | 2 | 5 | | | |
| 19 | Population cycles | | 1 | AT2 | | |
| 20 | Predator–prey relationships | | 1 | AT2 | | |
| 21 | Plant–herbivore interactions | | | | | |
| 22 | Island biogeography theory | | | | | |
| 23 | Bioaccumulation in food chains | 2 | 1 | AT2 | | |
| 24 | Co-evolution | | 4 | AT2 | | |
| 25 | Stochastic processes | | | | | |
| 26 | Natural disturbance | | | AT3 | | |
| 27 | Habitat restoration | | | | AT5 | * |
| 28 | The managed nature reserve | | 1 | AT2/3 | | * |
| 29 | Indicator organisms | | | | | |
| 30 | Competition and species exclusion | | | | | |
| 31 | Trophic levels | 9 | 4 | | | |
| 32 | Pattern | | | AT2/3 | | * |
| 33 | *r* and *K* selection | | | | | |
| 34 | Plant/animal co-evolution | | | AT2 | | |
| 35 | Diversity/stability hypothesis | | | | | |
| 36 | Socioecology | | | | | |
| 37 | Optimal foraging | | | | | |
| 38 | Parasite–host interactions | 2 | 3 | | | |
| 39 | Species–area relationships | | | | | |
| 40 | The ecotype | | | | | |
| 41 | Climax | 2 | 1 | | | |
| 42 | Territoriality | | | | | |
| 43 | Allocation theory | | | | | |
| 44 | Intrinsic regulation | | | | | |
| 45 | Pyramid of numbers | 3 | | AT2 | | |

Table 2.1. (*cont.*)

| BES rank | Principal concepts of ecology | GCSE | | Natl. Curr. | | |
|---|---|---|---|---|---|---|
| | | Sci. | Biol. | Sci. | Geog. | XCurr. |
| 46 | Keystone species | | | | | |
| 47 | The biome | | | AT2 | | |
| 48 | Species packing | | | | | |
| 49 | The 3/2 thinning law | | | | | |
| 50 | The guild | | | | | |

The concept list is after Cherrett (1989).
BES: British Ecological Society. GCSE: General Certificate of Secondary
Education, Sci: Science, Biol: Biology. Natl. Curr.: National Curriculum, Sci:
Statutory Orders for Science (1991), Geog: Statutory Orders for Geography
(1991), XCurr: Cross-curricular: Non-statutory Guidance Number 7:
Environmental Education (1990). AT: Attainment Targets.

Table 2.2. *Ecological and environmental issues included in the school
curriculum but not identified as central to ecology in Cherrett's survey of
practising ecologists*

| Ecological and environmental issues | GCSE | | Natl. Curr. | | |
|---|---|---|---|---|---|
| | Sci. | Biol | Sci. | Geog. | XCurr. |
| Pollution | 11 | 5 | AT2 | AT5 | * |
| Population size/control | 5 | 3 | | AT5 | |
| Soil (as a resource) | 7 | 3 | | AT3/5 | |
| Artificial ecosystems | | 6 | AT2 | AT2/6 | * |
| Resource exploitation | | | AT2 | AT5 | * |
| Ecological 'issues' | | | AT2 | AT1/5 | |
| Seasonality/seasonal changes | | | AT4 | AT5/2 | * |
| Colonisation | | 4 | | | * |
| Abiotic factors | | 2 | | AT5 | |
| Field techniques | | 4 | | AT1 | |
| Engender respect/attitude | | | AT2 | AT5 | * |
| Industry and environment | | | | AT2 | |
| Environmental management | | | | AT5 | |
| Environment and leisure | | | | AT5 | |
| Global environmental change | | | | AT5 | |
| Fragility of ecosystems | | | | AT5 | |
| Climate and vegetation | | | | AT5 | |
| Sustainable development, stewardship, conservation | | | | AT5 | |

BES: British Ecological Society. GCSE: General Certificate of Secondary
Education, Sci: Science, Biol: Biology. Natl. Curr.: National Curriculum, Sci:
Statutory Orders for Science (1991), Geog: Statutory Orders for Geography
(1991), XCurr: Cross-curricular: Non-statutory Guidance Number 7:
Environmental Education (1990). AT: Attainment Targets.

Reform Act may resolve some of these issues but, at the same time, may raise other problems.

Traditionally, education about the natural environment in schools has been included in programmes of study identified as nature study, rural studies, field studies and as components in geography and biology courses at most levels. These studies have largely described how the natural world functions. This is important but there are human factors which must not be ignored if students are to gain an understanding of the current issues which will help them to identify the changes necessary to achieve long term solutions.

It can be seen that the quality of environmental education in the United Kingdom has been inconsistent. However, over the past five years a number of developments have taken place that have raised the profile of environmental education in both the formal and non-formal (i.e., out of school time) education sectors.

One of the most significant of these has been the Resolution of the Council of Education Ministers of the European Community on Environmental Education, which emphasised 'the need to take concrete steps for the promotion of Environmental Education . . . throughout the Community'. As a result, the Education Ministers considered 'that as a matter of priority Environmental Education should be promoted within all schools of the Community' (EC Resolution 88/C177/03).

In the UK the role of environmental education in the school curriculum was described in 1989 by HM Inspectorate:

Schools have a role in helping their pupils make sense of their experiences and in developing their knowledge and understanding of the physical and human processes which interact to shape the environment. Schools can also help to foster a reasoned and sensitive concern for the quality of the environment and for the management of the earth's resources. These are, of course, matters of increasing social concern.
*(DES, 1989)*

The Government's White Paper on the environment, *This Common Inheritance*, published at the end of 1990 (DOE, 1990), devotes a chapter to 'Knowledge, education and training', in which it is acknowledged that environmental education is necessary to ensure the effective implementation of environmental policy through an aware and informed population.

However, in the formal education sector emphasis has usually been placed on disciplinary rigour which has failed to present students with coherent learning experiences. Additionally, ecology and environmental science have suffered from being considered a 'soft' science by many and have presented problems to those teachers who have little or no scientific background.

## The National Curriculum for schools in England and Wales

In passing the Education Reform Act, the UK government is for the first time setting the educational agenda in terms of what is to be taught and assessed. Other legislation, such as the local management of schools and charging for out of school activities, and the introduction of new examinations and schemes aimed at vocational training, has been the catalyst for these changes (Hale, 1990).

The National Curriculum Council, established under the Education Reform Act 1988, was responsible for the development and monitoring of the National Curriculum. The National Curriculum has been described as 'an instrument for change and improvement in the English educational system' which will 'raise standards and provide a new impetus by encouraging best practice' (NCC, 1990). Every state school is now required by law to provide a basic curriculum consisting of the subject areas defined by the National Curriculum plus religious education. The ten subjects are: English, mathematics, science, technology, geography, history, modern foreign languages (11–16 only), art, music and physical education.

The cross-curricular elements transcend the core and foundation subject boundaries and serve to unify the whole curriculum. The common threads in the curriculum are these cross-curricular dimensions, skills and themes.

Environmental education is one of the five 'cross-curricular themes'. The others are careers, economic and industrial understanding, health education and citizenship. Documentation and guidance for these give schools strategies to provide coherence and continuity. Environmental education is reflected in the content of some subjects (notably science and geography) and, because it is a cross-curricular theme, should be included in all other subjects, as both a starting point and a unifying element (Palmer, 1991).

Skills such as problem solving, study and communication skills, which are fundamental to environmental education, are central to all subjects of the National Curriculum. Environmental education is therefore ideally suited to be taught as a cross-curricular theme rather than as an individual subject.

Much of the pedagogy of environmental education depends on an integrated and cross-curricular approach. Thus, the conceptual base and teaching methods in environmental education are interdisciplinary and implemented through a topic approach.

### The National Curriculum subjects

The Education Reform Act (1988) requires Statutory Orders for each foundation subject to prescribe Attainment Targets and programmes of study for the ten foundation subjects.

The Attainment Targets and the programmes of study have been defined. The first phase of the Introduction of Attainment Targets and programmes of study in science and mathematics commenced in September 1989.

### The development of ecological and environmental elements in the National Curriculum

Increasing public concern for the quality of the environment, as well as resolutions of the European Parliament, have resulted in a sharp focus on ecology and environmental education so that access to a curriculum incorporating environmental issues is an entitlement for all.

It was a reflection of this increased concern that in September 1988 the then British Prime Minister (Margaret Thatcher), signalled the government's commitment to the environment by drawing attention to the interdependence between health, jobs, industry, the economy and the environment.

### Ecology in the National Curriculum

#### (i) The Science National Curriculum

The National Curriculum for Science details programmes of study and Attainment Targets for students from 5 to 16 years. Within these, investigative skills and communication assume major importance.

Practical experience is a vital component of ecological science and can meet the requirements of the science curriculum. Field observation and experimentation are central to ecological science, providing data on the nature of ecological systems and allowing predictive models to be tested.

Ecology has generally been regarded as a 'specialist' subject within biology (Hale, 1986). As a consequence, it has had a low profile in biology syllabuses. The Science National Curriculum contains approximately a 15% biological science component (Dunkerton and Lock, 1989). Of this, pure ecology can be said to comprise a minor component (see DES/Welsh Office, 1989).

The original National Curriculum for Science (1989) comprised 17 Attainment Targets (ATs). These have now been reduced to four (DES/Welsh Office, 1991) by the amalgamation of several of the original Attainment Targets into each of the 'new' ones. The content of the science curriculum has remained essentially the same.

Of the four Attainment Targets for science (revised Orders for Science, DES/Welsh Office, 1991), Attainment Target 2 ('Life and Living Processes') contains an ecological component via two of the four 'strands' of progression running through the curriculum from ages 5 to 14. These are 'Populations and human influences within ecosystems' and 'Energy flows and cycles of matter within ecosystems'.

At Key Stage 1 (age 5 to 7 years), for example, the programme of study stipulates:

Pupils should study plants and animals in a variety of local habitats, for example, playing fields, garden and pond. They should discuss how human activity produces local changes in their environment.

By the time Key Stage 3 (age 12 to 14 years) is reached:

Pupils should study a variety of habitats at first hand and make use of secondary sources, to investigate the range of seasonal and daily variation in physical factors, and the features of organisms which enable them to survive these changes. They should be introduced to the factors affecting the size of populations of organisms, including competition for resources and predation. They should study the effects of human activity, including food production and the exploitation of raw materials, on the purity of air and water and on the Earth's surface. They should come to appreciate that beneficial products and services need to be balanced against any harmful effects on the environment.
*(revised Orders for Science, DES/Welsh Office, 1991)*

However, the Science National Curriculum falls far short of providing an adequate and balanced approach to ecology, as key concepts (such as succession, niche, and competition) are omitted with issue-based topics being given greater prominence (see the preceding section 'The nature of ecology' and Table 2.1).

*(ii)  The Geography National Curriculum*

The Geography National Curriculum includes 'Environmental Geography', which provides pupils with the opportunity to study issue-based environmental topics and should thus complement the concept-based theoretical ecology in the science curriculum (Hale, 1991).

However, 'Environmental Geography' does have areas of overlap with the science curriculum. For example, in the programmes of study for Environmental Geography, at Key Stage 1, pupils should be taught:

To identify activities which have changed the environment and to consider ways in which they can improve their own environment, [also] how the extraction of natural resources affects environments.

At Key Stage 4 (age 15 to 16 years):

Pupils should be taught about fresh water sources and means of ensuring a reliable supply; why rivers, lakes and seas are vulnerable to pollution [and] ways in which people look after and improve the environment.

The similarities with science as detailed above are obvious.


*(iii)  Cross-curricular themes*

The five cross-curricular themes of the National Curriculum are required to be taught across all subjects at every Key Stage. All schools have a responsibility to deliver a coherent programme for environmental education which will enrich all aspects of the curriculum. Thus, the environmental context can give issues and problems raised in, say the science, music and history curricula, a relevant and topical perspective.

The National Curriculum Council publishes help for teachers in the form of non-statutory guidance; that for environmental education was published in Autumn 1990 as *Curriculum Guidance No. 7: Environmental Education* (CG7). This provides a rationale for environmental education in terms of both educational and environmental needs and sets out a framework within which the aims and objectives of environmental education are placed in the form of the purpose and means of achieving an environmentally literate population (Harris, 1990). It advocates a whole school approach to both the formal and hidden curricula and the planning of environmental education across the whole curriculum using the 'about, for, in and through the environment' model. Much of CG7 is devoted to describing case studies of good practice in environmental education in schools.

According to the National Curriculum Council the three main aims of environmental education are to:

1. Provide opportunities to acquire the knowledge, values, attitudes, commitment and skills needed to protect and improve the environment.
2. Encourage pupils to examine and interpret the environment from a variety of perspectives – physical, geographical, biological, sociological, economic, political, technological, historical, aesthetic, ethical and spiritual.
3. Arouse pupils' awareness and curiosity about the environment and encourage active participation in resolving environmental problems.
(*Curriculum Guidance 7*, 1990, page 3.)

Environmental education may be taught in a variety of ways in a wide range of contexts. Environmental work may take a number of forms, such as study in the local surroundings, the understanding of natural ecosystems, the development of conservation areas, the design of practical projects with the local community or investigation of global issues such as endangered species and tropical deforestation (Dorion, 1994).

The cross-curricular requirements of the National Curriculum help to reinforce the fact that subject disciplines are not discrete and that there is overlap and interaction between them. Thus, ecological work may be included in aspects of history, modern foreign languages, mathematics, English, etc., as well as in biology and geography. For example, past types and levels of pollution may be studied in a range of countries using reports and articles in different languages and compared with contemporary aspects of environmental quality.

## Ecology teaching in schools

In general in primary schools (5 to 11 years) greater use of the environment has been made for teaching. However, the quality and extent of this approach has been highly variable as no statutory requirements or guidelines have existed nationally prior to the National Curriculum.

The planning of schemes of work and curriculum audit in secondary (11–16/18) schools should identify and allow for complementary and related work in different subject areas to be organised in a logical framework. For example, the science curriculum will provide the basic understanding of ecological concepts and processes and geographical studies will allow for the application and extension of the skills, knowledge and understanding of ecology in particular case studies and investigations (Hale, 1991).

While it is encouraging that practical techniques of ecological investigation are extended to develop analysis, knowledge and understanding of ecological processes and for acquiring relevant skills, there are a number of logistical problems in implementing these recommendations in schools:

* the absence of readily accessible habitats;
* lack of sufficient time in the daily timetable to undertake locally based field work;
* large class sizes, making assessment difficult;
* lack of sufficient resources, including equipment and materials, etc.

## Training

One of the most recurrent identified needs for the past 20 years has been to train teachers in environmental education. It is implicit in the Brundtland Report, which emphasises the need for public education in environmental matters. UNESCO–UNEP (1990) indicated the crucial need for the 'environmentally educated teacher' as the role of environmental education is fundamental to the care and protection of the environment. The recent government proposals for changes to teacher training do not indicate sufficient improvement.

## Conclusion

Ecology is concerned with the application of ecological knowledge for the more effective management of the environment. To this end there is a need to raise ecological awareness generally and to promote a sound ecological base in education at all levels. There is now a real opportunity within the framework of the National Curriculum for ecology and environmental education to be part of every pupil's basic education. A sound and well researched entitlement to environmental education is now feasible, integrated within the statutory guidelines of the National Curriculum.

Ecology is a recognised component of the Attainment Targets that comprise the science curriculum and, together with components in the geography curriculum and the cross-curricular themes, can be taught in an integrated and meaningful way.

The inclusion of ecology in the curriculum draws on and fosters a range of basic and transferable skills and serves many of the objectives of the National Curriculum. In addition, ecology has a fundamental contribution to make to process based science education and allows for a broad understanding of global and local issues (Hale, 1988, 1991).

A sound and balanced ecological understanding is a prerequisite for a meaningful analysis and evaluation of topical environmental issues. Cherrett's survey conclusions provide a clear guide to the fundamental principles of ecology which curriculum developers, teachers and those involved in environmental training should incorporate in their courses. However, as this is a new area of study for many schools there is a need for teacher guidance and training as well as additional resources and support.

The World Conservation Strategy (IUCN/UNEP, 1980) put forward the view that 'the expansion of environmental education was an essential instrument to changing the behaviour of entire societies towards the biosphere'.

As Martin (1990) emphasised, there needs to be a strategy for environmental education, and for it to be effective, each of the elements which contributes to education, including formal education (schools and colleges), the media, socio/cultural influences, such as religious networks, and professional and occupational training, must be responsible for delivering environmental education as an aspect of its activities.

## References

Cherrett, J.M. (1989). Key concepts: the results of a survey of our members' opinions. In *Ecological Concepts*, ed. J.M. Cherrett, pp. 1–16, Oxford: Blackwell.

Department of Education and Science and the Welsh Office (1989). *Science in the National Curriculum*. London: HMSO.

Department of Education and Science and the Welsh Office (1991). *Science in the National Curriculum Revised Order for Science (1991)*. London: HMSO.

Department of the Environment (1990). *This Common Inheritance*. London: HMSO.

Dorion, C. (1994). *Planning and Evaluation of Environmental Education 11 to 16*. UK: World-Wide Fund for Nature.

Dunkerton, J. and Lock, R. (1989). Biological Science: equal opportunities and choice. *Journal of Biological Education*, 23 (3), 164–165.

Hale, M. (1986). Approaches to ecological teaching: the educational potential of the local environment. *Journal of Biological Education*, 20 (3), 179–184.

Hale, M. (1988). *Ecology in the Curriculum: 5–19*. London: The British Ecological Society.

Hale, M. (1990). Recent developments in environmental education in Britain. *Australian Journal of Environmental Education*, 6, 29–44.

Hale, M. (1991). Ecology in the National Curriculum. *Journal of Biological Education*, **25** (1), 20–26.

Harris, C. (1990). *Curriculum Guidance Number 7: Environmental Education*. CEE Briefing, Council for Environmental Education.

International Union for the Conservation of Nature and Natural Resources (IUCN), with the United Nations Environment Programme (UNEP) and the World Wildlife Fund (WWF) (1980). *The World Conservation Strategy*. Gland, Switzerland.

Martin, P. (1990). *First Steps to Sustainability: The School Curriculum and the Environment*. UK: World-Wide Fund for Nature.

National Curriculum Council (1990). *Curriculum Guidance Number 7: Environmental Education*. London: HMSO.

Palmer, J.A. (1991). Implementing CG7: Policy into Practice. *Environmental Education*, **36**, 9–10.

UNESCO-UNEP (1990). *Environmentally Educated Teacher*, Connect, Vol. XV(1).

World Commission on Environment and Development (1987). *Our Common Future*. Oxford: Oxford University Press.

# 3

# Environmental education in the primary school science curriculum in Uganda

CHRISTOPHER ACAR

*Teaching Assistant, Institute of Teacher Education, Kyambogo, (ITEK), Uganda*

## Introduction

The concept of environmental education is not a new one in the primary school curriculum in Uganda. Before the 1970s environmental and health education were taught in schools as nature study and hygiene. Today, environmental education is integrated into the science curriculum, which is one of the core subjects at the primary level.

The history of environmental education in schools in Uganda can be traced back to the mid-1980s, when the Ministry of Environmental Protection was established to monitor, mobilize and sensitize people about, fundamental issues of environmental degradation. This important development has its origins in the political and social situation at the time. After nearly two decades of political instability, many systems in the country, especially social services, had seriously deteriorated, if not completely collapsed. Notable deterioration had occurred especially in the areas of health and education.

In 1987, a major review of the primary school science syllabus was undertaken jointly by the National Curriculum Development Centre (NCDC) and the Inter-Ministerial Advisory Panel (IMAP), (formerly Inter-Ministerial Expert Panel – IMEP). The members who constituted the IMAP were drawn from the various ministries including: Education, Agriculture, Health, and Environmental Protection as well as from non-governmental organizations (NGOs), such as UNICEF and UNESCO. The composition of the IMAP also indicates the significance that is now beginning to be attached to participatory planning in education.

Since 1987, environmental and health education has become a major element of primary school curricula. The primary science panel of the NCDC has adopted the policy that science education should draw extensively on everyday experiences of pupils and prepare them for adult and working life.

The New Basic Science and Health Education syllabus for primary schools in Uganda aims among other objectives to:

(i)   stimulate interest in and care for the environment
      [and]
(ii)  promote awareness that application of science may be both beneficial and detrimental to the individual community and the environment
*(NCDC, 1987: pp. 12–13)*

The science course at primary level aims to help young people understand environmental problems and also give them the opportunity to acquire knowledge and skills that will encourage them to become involved in environmental control, protection and development. It also aims to enable them to analyze and evaluate their relationships with other people and their surroundings. This is achieved through understanding the ecological processes that govern life in the environment in which the pupils live.

For many rural children in Uganda, primary education may be the only education they can afford. Many children are forced to leave the school system after completing the primary phase. Large families are common among the rural poor and school fees at the secondary level of education can be found for only one child in the family, the other children then have to leave school and engage in productive and economic activities to pay the fees of the one who remains in education (who is usually a boy).

Primary education in Uganda is therefore the most significant means of disseminating basic and essential ideas for community development and transformation. In this respect children have a special role to play.

In Uganda, as in most developing nations, children participate actively in agriculture: hoeing, sowing seeds, weeding, harvesting, threshing and storing the crops. They are also engaged in pest control. Customarily, there is a distribution of labour between males and females in Uganda. Girls collect firewood and undertake cooking, while boys tend the animals and build homes. Sometimes children from poorer homes are hired out by their parents to work in fields of more well-to-do families for small payments.

By carrying out these activities children make a contribution to earning income and providing food for the family. Many environmental problems in the community are part of the children's daily experiences. Making provisions for environmental education in the primary school curriculum gives children the opportunity to reflect and focus on their own problems in their homes and villages. Through these lessons children are expected to analyze the realities of their situations and to learn to appreciate the natural environment and the ecological conditions in which they live.

This paper will explore the content, relevant approaches and constraints to the implementation of the environmental and health education programmes in the primary school science curriculum in Uganda.

**Global environmental issues**

Deforestation, pollution, desertification, land degradation, disposal of toxic and hazardous waste and the effects of acid rain have been recognized as environmental problems on a world scale over the last two decades. However, since the 1980s the world has seen the emergence of additional environmental challenges such as global warming, the depletion of the ozone layer and the HIV/AIDS disease.

The resolution of these issues is fundamental to the future well-being of this planet and of its inhabitants. Due to the rapid and intensive changes occurring in the environment, a new impetus for environmental awareness is also developing on a worldwide scale. Individual choice and use of the environment are influenced through the development of insights, awareness, knowledge, skills and positive attitudes, the essential competencies that enable people to make reasonable responses. These competencies are central attributes necessary for both intellectual and practical decision making. They provide a full understanding of actions that will result in a reduction in the damage to the environment and adverse effects to society. This ultimately leads to an improvement in the quality of life for all people in the world.

In Uganda, the primary school system is recognized as one of the most effective means of conveying messages of environmental awareness. This realization stems from the fact that primary schools are numerous and widespread and that the majority of the country's population is within the primary school age group (i.e. 6–12 years). As future parents and leaders of the nation, they will become powerful instruments of change.

**Environmental education issues in Uganda**

The nature of environmental problems in the developed world is different from that in developing nations. These, in turn, vary from place to place. In Uganda environmental education places emphasis on local issues such as the burning of bush fires, over-fishing, over-grazing, over-cultivation, deforestation, soil erosion, population control, dumping waste in water and the 'poaching' of animals from National Game Reserves. These problems are interwoven into the cultural, social and economic lives of rural communities. It is therefore difficult to disentangle environmental problems from the needs and interests of a society where 85–90% of the entire population rely heavily on subsistence agriculture.

Many people in these communities are not sensitized and fully aware of the dangers posed by environmental problems and therefore do not regard them as immediate risks. A number of animal and plant species are on the verge of extinction or are already extinct due to a lack of awareness and appropriate action.

For example, hyenas are no longer found in Uganda. About three decades ago, hyenas used to inhabit the grasslands in the rural countryside in packs, particularly in northern Uganda, hunting and searching for carrion. Occasionally, the hyenas would attack farmers' sheep and goats. Sheep and goats are one of the most treasured assets of the rural farmer. Farmers eradicated the hyenas in this area of northern Uganda. However, a way should have been found for the farmers to co-exist with the hyenas.

Two points arise out of this. Firstly, environmental education contradicts the interests of some people, and secondly, in the developing world there are limited means to deal with such conflicts and seek satisfactory resolutions.

Similar examples of indigenous species control can also be cited. One is that of elephants. These can be very destructive to villages as they eat crops, which are the only means of livelihood of peasant farmers. Similarly, swarms of weaver birds also cause severe damage to farmers' grain (millet, sorghum and maize) during the harvest season.

Many serious environmental problems are not easily resolved by the rural communities themselves and are simply left unattended to, with the result that they are likely to intensify.

One of the methods adopted in Uganda in an effort to resolve environmental problems and to develop environmental awareness among the population has been to mobilize children in primary schools as agents of change. The programmes in the schools are viewed as long-term strategies.

The primary school syllabus currently recommends a wide variety of teaching approaches. These emphasize practical investigative and problem-solving activities based on children's immediate environment, rather than the accumulation of scientific knowledge. The activities are graded to ensure a progressive development of attitudes and skills, which are vital in the promotion of environmental protection.

Considerable emphasis is also being placed on the application of scientific principles to real world situations through activities such as seed planting, tree planting using indigenous tree species from children's and the community's own reared saplings, biological control of pests and filling gullies resulting from soil erosion.

Most of these activities are co-ordinated through a national network of programmes developed under the Namutamba Project of Basic Education Integrated in Rural Development (BEIRD). Based at Namutamba Teachers' College, this programme was intended to spread and operate in all the primary schools and Primary Teachers' Colleges (PTC) in the country.

Another programme, the 'MINDSACROSS' currently being developed at the Institute of Teacher Education Kyambogo is a teacher education programme and is intended to be disseminated to all Primary Teachers' Colleges. It comprises activities that integrate school and community life. The programme also seeks to involve children in the production of their own reading materials at the lowest cost. This innovative project has been initiated by the Institute of Teacher Education Kyambogo with the financial assistance of the International Development Research Centre (IDRC), Ottawa.

'MINDSACROSS' believes that it is possible to produce relevant indigenous literature on major societal issues through processes that motivate schools and their communities to examine and change for the better the conditions in which they live.

The central idea underlying the 'MINDSACROSS' project is that those who need books covering issues of major concern must produce them themselves, so bypassing professional publishing, which is difficult for schools and communities.

In Uganda environmental education is considered a major part of the curriculum, as it has a direct bearing on the present and future lives of children. Hence, it is an important component of the 'MINDSACROSS' project.

As part of the project, non-literate adults can recount stories and histories or identify problems of major concern in the community. This can be taped and transcribed. The aim is to encourage school children and adults to feel that they have something worthwhile to say and to create a network of teachers, students, children and adults that form a community of writers, readers and communicators. Through this exchange of ideas the community can address problems and issues in both school and community.

In one of the pilot colleges, Naziggo Teachers' College, together with five nearby primary schools affiliated to it, the 'MINDSACROSS' project is taking root rapidly and successfully.

Innovative teachers are being encouraged to write units and activities that will actively involve pupils. They are also encouraged to carry out basic research that will enable them to assess and evaluate the success of environmental education programmes in the school and its locality. For instance, the unit of the primary school syllabus 'our environment' is taught across the infant range (i.e. 6–8 years). This unit enables children to explore their immediate environment and become more aware of the factors that may influence their lives within the school environment.

While learning about 'our environment', children are exposed to a wide variety of living and non-living components of their everyday lives. At this stage children are given the opportunity to discuss the value of the components they have observed and how these affect each other in the environment (interdependence).

Older children are encouraged to expand their knowledge to national and global environmental issues by exposure to a variety of sources of information such as environmental talks, stories, magazines, radio talks, and newspaper cuttings, as well as by role play and simulation activities. Stories or talks by the teacher or an outside person (for example, someone working for the promotion of environmental protection) are arranged for the pupils. A recorded radio talk (in the vernacular language) may also be used to assist and challenge both the children and students at teachers' colleges with environmental problems with which they are already familiar.

A number of environmental activities are found in the regional children's magazine, *The Rainbow*, which is produced and published by The Mazingira Institute, Nairobi and distributed free to primary schools in East Africa. In practice not all the aims of the project are successfully being implemented in schools.

The Education Policy Review Commission (1989) recommends that primary education should lead to the acquisition of:

(i) knowledge of the natural environment and its utilization and conservation of physical and biological science;
    [and]

(ii) knowledge of the social environment and social institution, civic rights and responsibilities, the country's culture, history and geographical features
*(Education Policy Review Commission, 1989: pp. 43)*

The Commission also recognizes that limited progress can be achieved when a large number of primary school pupils are not being able to continue to secondary education. It therefore recommends that a high priority should be given to primary education.

The United Nations Social and Economic Council (UNSEC), (1989), highlighting the problem of deforestation, states that:

In Uganda 50 000 hectares of forest land are cut each year

The purpose of cutting down trees in Uganda is mainly to release land for cultivation and for burning for charcoal fuel. At the present rate large areas of land are becoming bare each year. Pupils in primary schools participate actively in these destructive activities, yet at school they are taught about the dangers of cutting down trees and of not replanting and the effects this has on the environment. These lessons encourage pupils to meditate on the dangers of cutting down trees and to make conscious decisions about solving environmental problems such as this, which affect their lives and the general welfare of the future generations. This is the central theme of the environmental education programme in schools in Uganda.

One of the main difficulties in solving environmental problems in Uganda today is the failure of individuals to consider environmental values in favour of short term economic gains. This is clearly evident in the case of the problems caused by over-fishing and deforestation. In addition, there is a lack of awareness that it is possible to reconcile the needs for development with the needs of conservation. The link between ecology and sustainable economic growth needs to be promoted vigorously. The education provided in primary schools can make an effective contribution to this. Development can only proceed successfully within a sustainable regime.

Pupils in primary schools have a significant role to play in promoting basic concepts of sustainable development in their communities. The kind of environmental education programme being provided for pupils must be relevant to societal needs of the community. As a starting point pupils need to understand the nature of interdependence among living organisms; this will lead to a realization of the importance of conservation.

In support of this, Masterton (1987) commented that:

the most effective way to educate *for* the environment in all its individual, personal and natural, ecological, social and cultural dimensions is to work *in* and *through* in order first of all to learn *about* it.

This statement supports the emphasis being placed on the role of environmental education in providing cognitive understanding as well as development of the necessary skills and attitudes leading to action in the conservation and sustainable development of the environment.

Uganda is one of the many developing countries finding difficulties in repaying substantial external debts. In addition, poor management of resources, and numerous environmental challenges, coupled with the falling prices of Uganda's raw materials (notably coffee), has led to the exacerbation of these problems with no immediate resolution in sight. The implementation of structural adjustment policies has also taken a considerable toll on the rural poor.

As a consequence of these problems, development funds for education and environmental protection are not usually forthcoming, as these areas are not regarded by the government as priorities. While other sectors such as agriculture and defence consume 22% and 29% respectively of the total National Budget (1989/90) the burden of funding environmental protection is left almost entirely to non-governmental organizations and other voluntary organizations.

Carry American Relief Everywhere (CARE), an NGO, is actively involved in environmental protection in Uganda. It is currently sponsoring a US$35 000 project being undertaken by the Uganda Sigiri Industry Kampala (USIKA). This project aims to improve the design of the charcoal stove (the Sigiri), to limit the amount of fuel wood it burns and to maximize its efficiency.

USIKA have calculated that up to 35 000 tonnes of wood could be saved each year if 12 000 re-designed and more efficient stoves were installed. Other NGOs actively involved in helping with environmental protection in Uganda are the Uganda Red Cross, the Uganda Wildlife Club, the Young Farmers Club, the Danish Agency for International Development (DANIDA) and the United Nations Organization. These organizations are attempting to initiate and help the Ugandan people to implement environmental programmes in communities, with projects such as raising seedlings of trees from the indigenous tree species and planting them, and caring for and protecting some endangered species of both plants and animals.

## Environmental health

In many developing countries, preventable disease still claims millions of lives each year. The most vulnerable to these diseases are children within the primary school age range.

In Uganda the most serious of these diseases are malaria and diarrhoea. Many of the illnesses and deaths resulting from these diseases could be avoided if families, community leaders and in particular primary school teachers were adequately informed about matters relating to health.

*The State of the World's Children* (Grant, 1990) shows that the mortality rate of children under five in Uganda is currently 169 per 1000 live born and that only 18% of the rural population have access to clean and safe drinking water and adequate sanitation. Fifty-seven per cent have access to health services (the figures are slightly higher in each case for urban populations). The country's population is rapidly expanding at the rate of 3.4% per annum. A large population will exacerbate the situation, leading to increases in poor housing and overcrowding. Malnutrition is also a problem in some parts of the country. These problems call for positive action to be concentrated in these areas of need.

The contribution of primary education in preparing children for life in society has been recognized. The National Curriculum Development Centre (NCDC) in conjunction with the Inter-Ministerial Advisory Panel (IMAP), with the financial support of UNICEF, reviewed the primary science syllabus in 1987 (National Curriculum Development Centre, 1987). The revised syllabus, launched in May 1988, includes greater emphasis on health education. Of the 19 topics in the revised syllabus, nine are related to health education – i.e., 50% of the primary science curriculum and 12% of all teaching time is now devoted to health education.

There are nine health-related topics in the revised syllabus. These are: water and sanitation, AIDS control, food and nutrition, immunization, common diseases, primary health care, (PHC), our health, family health and social problems, and accidents and first aid. The syllabus aims to place science topics in context through an environmental approach to teaching and learning. Particular attention is given to the relationship between the knowledge and skills that children are expected to acquire, what they already know and the relevance of this to the lives of the children in the community.

Thus, the environmental context of primary education in Uganda serves the needs of science education by providing opportunities for the development of practical and intellectual skills that are relevant to the needs of society. Health education is being used to provide the vitally needed education in and for the environment.

In Uganda, health education practice has changed over recent decades. In the past health education in the community was aimed only at adults. Moreover, there was little community participation. Health education was carried out by health workers who were not trained as health educators. Today, the emphasis on health education has shifted from curative to a preventative approach. There is an increased community participation and pupils in primary schools have become 'health messengers', introducing new ideas about health to their friends, families and communities. Children are also involved in practical activities such as destroying the breeding places of mosquitoes in and around their school and home environments.

The work of primary schools in promoting the health of the nation is being enhanced and supplemented by support from the School Health Education Project (SHEP) and Health Education Network (HEN).

SHEP and HEN are health education promotion organizations operating under the umbrellas of the ministries of Education and Health respectively. Both organizations are funded by UNICEF. The team members of SHEP and HEN are predominantly trained health educators and their task is to organize and preside over health education conferences, seminars, symposia and meetings in schools and in the community. They also monitor and evaluate the success of health programmes and campaigns in schools and communities. Both organizations work in conjunction with their ministry officials, local government officials and community leaders. They also work hand in hand with teachers and other field officers.

Through the recommendations of SHEP and HEN local government officials may enact 'by-laws' for observation within the school and local communities. The role of primary schools together with the input of SHEP and HEN have already gone a long way towards the promotion of health and well-being among school children and community members.

As primary school pupils are the future leaders of the nation and the parents of tomorrow their knowledge, skills and attitudes are of crucial influence on the health and the environment of future generations. At Primary Teachers' Colleges, corresponding programmes on health and environmental education are given due emphasis. At the training colleges, teachers are being trained for both classroom teaching and community development. Even though the status and the morale of school teachers has been eroded over the past few decades (primarily due to economic hardship), teachers, especially in rural environments, are still respected members of their communities. Their example and advice may help to inform communities about simple low cost methods of preventing diseases and promoting environmental health. For instance, all young people who have completed primary education in recent years are now able to make a simple oral re-hydration solution, using water, salt and sugar in the correct proportions, to administer to children suffering from diarrhoea. This may save many lives in the future.

Also the fundamental message that one's health is one's own responsibility places pupils in a position where they can participate actively in making effective contributions to the health of the family, the community and the entire nation.

## Conclusion

Environmental and health education programmes in primary schools in Uganda combine formally planned and informal learning experiences. These contribute to the development of the appropriate knowledge, understanding, skills, attitudes and values on which individuals can base their choices and decisions relevant to their own environmental health and that of the community in which they live.

The success of the programmes outlined above necessitates co-operative and closer links between local communities and schools. The importance of these links has already been demonstrated by the operations of the Inter-Ministerial Advisory Panel (IMAP). With increased community participation, and willingness of pupils, teachers and parents, the environmental and health education programmes in primary schools will be of immediate benefit. However, much still remains to be done in order to implement the programmes fully. Judging by the results so far, the environmental awareness of pupils in primary schools has been heightened and has led to positive and caring attitudes towards their environment.

## References

Education Policy Review Commission (EPRC) (1989). *Education for National Integration and Development*. Kampala: Ministry of Education.
Grant, J.P. (1990). *The State of the World's Children*, UNICEF.
Masterton, T.H. (1987). *Environmental Studies: A Concentric Approach*. Oliver and Boyd.
National Curriculum Development Centre, NCDC/UNICEF/UNESCO (1987).

*The New Basic Science and Health Education for Primary Schools in Uganda*. Kampala: Ministry of Education.
United Nations Social and Economic Council, (UNSEC) – Hunger Project (1989). *The African Farmer*, **Dec.**, No. 2.

## Further reading

Baez, A.V., Knamiller, G.W. and Smyth, J.C. (1987). *The Environment and Science and Technology Education*. Pergamon Press.
Carnegie, R. and Hawes, H. (eds) (1989). *Child-to-Child and the Growth and Development of Young Children*. A report and resource book from International Seminar held at Nyeri, Kenya from 27th November to 1 December 1989. Nairobi: Kenya Institute of Education.
Department of Education and Science (1981). *Environmental Education – A Review*. London: HMSO.
Fyson, N.L. (1984). *The Development Puzzle*, 7th edn. London: Hodder and Stoughton/Centre for World Development Education (CWDE).
Hunt, A. (1988). Environmental education in the science curriculum. In *A Common Purpose: Environmental Education and the School Curriculum*. Worldwide Fund for Nature, WWF.
UNEP/UNICEF (1990). *The State of the Environment*.

# 4

# The place of ecology in adult education

PETER A. THOMAS

*Lecturer, Department of Adult and Continuing Education, Keele University, Staffordshire ST5 5BG, UK*

## Introduction

In Britain there is a long tradition of non-vocational courses for the general public. Several decades ago ecology courses were popular, but fashions have now changed and the word ecology has been confused with environmentalism and politics, reducing its appeal to the public. Yet with the current emphasis on tackling conservation at the grass-roots level, and increasing concern about green issues, there is a continuing need to develop in people an understanding and awareness of local, national and international ecology. Two particular needs are recognised:

1. to culture an understanding of how local ecosystems work and a better appreciation of the need for protection and enhancement;
2. to relate global ecological problems to their effects on the local area (making them more meaningful and real).

These needs are being met in Britain in two main ways. Firstly, by injecting ecology into traditional natural history courses, as both indoor and field activities, leading ultimately to completely field-based courses, including some overseas. Secondly, by new initiatives in courses for the general public in practical habitat management based on historical and theoretical perspectives. Emphasis has been placed on exciting curiosity and encouraging self-exploration of a problem aided by explanations of relevant theory.

## What is adult education?

Adult education in the context of this chapter implies non-vocational courses run for the general public which have no entry requirements and no examinations or qualifications gained on completion of the course.

Table 4.1. *Results of a survey of extra-mural students in Liverpool, England conducted in 1979–80 giving (a) educational background, and (b) reason for attending classes*

| (a) | Qualifications | Male (%) | Female (%) |
|-----|----------------|----------|------------|
|     | Professional[1] | 65 | 60 |
|     | Other | 35 | 40 |

| (b) | Reason for attending | % |
|-----|----------------------|---|
|     | Leisure | 78 |
|     | Employment | 12 |
|     | Both | 10 |

[1] Degree or other professional qualification.
Data modified from Bland (1981).

Such non-vocational courses are usually referred to as 'extra-mural' or 'liberal adult education' courses. Traditionally they have been organised by voluntary self-help groups such as The Workers' Education Authority (WEA) and, more recently, by universities (see Universities Council of Adult Education, 1972, and Wiltshire, 1983, for further information). Extra-mural courses are undertaken primarily for leisure; a study of a typical group of adult education students (Table 4.1) showed that some 78% attended courses purely for recreation and only 12% came solely for reasons of their employment (these were mainly social workers attending relevant courses). A corollary is that, unlike in schools or higher education, participants are not galvanised into coming to an unpopular course by the threat of examinations; most people come only if they are interested in the subject.

## Ecology and adult education in the past

Undercurrents of ecology surfaced in amateur natural history societies in the early part of the century when the works of Warming and Clements reached our shores from the United States. This informal 'adult education' arising from within the community led to the popularisation of ecology from the early days of the science. Later, the work of Sir Arthur Tansley and his book *The British Islands and their Vegetation* (1939) and his more popular *Britain's Green Mantle* (1949) led to a flood of adult education courses in ecology in the late 1940s and 1950s. These were often remarkably advanced, with students tackling concepts at the leading edge of the developing science.

It is difficult to obtain accurate data on the number of ecology courses offered in Britain between then and now. The number of extra-mural 'biology' courses in Britain has risen from 723 per year over the period 1971–76 to 876 per year over the period 1982–87, an increase of 21% (data from the Universities Council for Adult and Continuing Education (UCACE) annual reports). However, the picture is not as positive as it may seem from these figures because the total number of extra-mural courses increased over the same period by 41%. Moreover, courses dealing purely with ecology have consistently declined over the last few decades (this is substantiated by word of mouth; it is difficult to get accurate numbers, partly due to the variety of ambiguous course names used in the records of individual universities and partly due to the broad course categories used each year by the UCACE). The overall impression is that ecology courses have declined from their azimuth in the 1950s to become a small element in the range of extra-mural courses now offered in Britain. This is not a deliberate policy decision on the part of universities; it is essentially governed by the interest of the public. If courses are not well attended they are less likely to be offered in subsequent years and will eventually fade from the programme. In the author's experience, an extra-mural course with the word 'ecology' in the title is severely handicapped and will probably fail to recruit sufficient participants to be viable.

The reasons why ecology courses are no longer popular are not easy to identify but there appear to be two main underlying reasons. Firstly, with the advent of mathematical modelling, computer analysis and seemingly ethereal concepts (all part of the science coming of age), ecology is increasingly perceived as complex, dull and incomprehensible to ordinary people, with little relevance to what is observed when looking at the natural, semi-natural and cultivated landscape. The original sparkle of 'scientific natural history' or 'interpreting the landscape with common sense', which was so much a part of the initial appeal of ecology, is being lost. As a seasoned course participant recently commented to the author, 'In my day an ecologist looked at a woodland with wise eyes and could explain to me what I was looking at, now he seems to stare at a computer screen with wide eyes and can tell me nothing that is meaningful'. While many of us would dispute this idea it nevertheless nicely illustrates the public's perception of ecological science. Secondly, the notion of ecology as a science dealing with the workings of the living world has been blurred to include green politics and environmentalism (Gronemeyer, 1987).

So, in the mind of the course-going public, a course on the 'ecology of national parks' is likely to conjure up thoughts of conflicts over land-use, moral dilemmas over their preservation, and pressure groups. Surprisingly in these green-conscious times, courses on environmental problems, whether local, national or international, appeal only to the dedicated few. The reason appears to be that people are unwilling to use their leisure time for what are perceived as depressing and emotionally demanding subjects such as acid rain, global warming and the problems of national parks.

Thus the pendulum of interest in the natural environment has moved away from ecology back to the traditional area of natural history. However, the author and many others in university adult education maintain that ecology still has an important place in courses designed for the general public.

## Why is ecology of value in adult education?

Encouraging ecological knowledge in the general public is of value from three points of view.

1. A general understanding of why plants and animals look and live the way they do based on ecological principles, and consequently why plant communities take on a characteristic appearance, increases enjoyment of outdoor leisure pursuits and hence improves the individual's quality of life; a respectable end in itself. Ecological knowledge and understanding also fosters concern for the protection of natural and semi-natural areas, an important consideration in Britain with its high population density and pressure on the countryside.

2. More specifically on a local scale, the current emphasis in Britain is on tackling conservation at the grass-roots level. People in communities are encouraged to have an interest in conservation on the assumption that this will be far more effective than dictates from local and national government. A key to its effectiveness is encouraging motivation and developing relevant skills within the community. This is where adult education classes play an important role. From the study summarised in Table 4.1, for example, it is clear that more than half the participants on courses have a degree or professional qualification and these people as a whole tend to be well informed about current affairs and are likely to be most involved and vocal in local affairs, including conservation (Feltwell, 1983). These people, then, are the very ones who would benefit from an understanding of how local ecosystems work and a better appreciation of how they can be protected and enhanced.

As an example, a local farmer wished to remove a small area of woodland, arguing that it was of little value since it was a mixture of dead elms (*Ulmus* species) killed by Dutch elm disease, and sycamore (*Acer pseudoplatanus*), an introduced and invasive weed. Local residents, however, noted that the woodland contained plant species indicative of an ancient woodland (their interest and information arising from attending an adult education class). The residents objected to the potential removal of the woodland on the grounds that despite its present appearance it was a relict of a much older woodland now rare in Britain. Detailed ecological surveys revealed several uncommon insect and plant species. In the light of these discoveries the farmer readily agreed to a management plan produced with the help of the local Wildlife Trust and implemented with voluntary help from the community.

3. On a national and international scale, there is mounting public concern about green and global issues such as damage to the ozone layer and deforestation. It is often argued that public pressure is important in persuading governments to change policies on, for example, energy production, and in prompting individual action such as insulating homes or taking domestic refrigerators to have the chlorofluorocarbons (CFCs) drained for proper disposal. Global problems, however, are often regarded by the public as being remote, and of someone else's concern. Adult education classes are valuable in assessing and discussing what is fact and what rhetoric, and also in relating global ecological problems to their effects on the local area, making them more meaningful and real. On a recent course on the national parks of central England the tutor and participants were discussing human impact on the natural vegetation and the question of global warming was raised. This was quickly dismissed by some participants as being of little relevance or importance. But using predictions based on current data of likely changes in the climate that may occur over the next 50 years, the class was able to piece together a likely scenario of changes that could be expected. This included problems of trees producing leaves earlier in spring with a consequent greater chance of frost damage to flowers leading to lower fruit production; repercussions for herbivores and long-term effects on populations; also, the paradoxical delay in spring flowering of plants with bulbs due to insufficient vernalisation and the resulting problems for woodland herbs that need to complete their growth before the leaf canopy closes. Ideas from many areas of ecology were used to piece together the scenario; ideas that discussed on their own would have stimulated little interest.

Political and social education has always been an important function of adult education (Jones, 1986) and indeed, much of the discussion above is primarily concerned with aesthetic enjoyment and with conservation and environmental concerns. Nevertheless, it should be emphasised that such end points are based on a sound understanding of the science of ecology.

In summary, ecology does have an important place in the education of adults. But if people are not attracted to an ecology course named as such, how then are adults to be encouraged to enrol for these courses? It is often a case of trying to provide people with what they do not think they want or can understand but really do and can! In schools and higher education, students can be forced to study ecology as a compulsory part of many syllabuses, although this does not necessarily mean that ecology will be taught well or enthusiastically by motivated teachers. With liberal adult education, however, ecology has to be 'sold' as an interesting subject to each student or, since they are paying to attend in their spare time, they may not return for future classes. How is this to be achieved? The solution is illustrated in the popularity of television nature documentaries, which have retained the image of 'scientific natural history' in bringing the vast world of science to the public in a form that is enjoyable, interesting, meaningful and digestible. At Keele and other universities in Britain similar approaches are being followed in two ways, firstly, by injecting ecology into traditional natural history courses and, secondly, by developing courses in practical habitat management.

**Natural history courses**

Experience has shown that many people are attracted to courses by phrases and words in the course title such as 'wild flowers', 'trees' and 'identification'; course titles incorporating these magic ingredients usually recruit well. Once people are safely in the class and have been made to feel comfortable, a chain is developed. The obvious starting point is the identification of plants (the 'what' and 'how' questions: 'what have I got here and how can I tell it from everything else?'); this, after all, is what most people have enrolled on the course for. Identification generally leads to enquiry as to where different species can be found (the 'where' questions) which in turn can be used to stimulate an interest in 'why'. Alternatively, the theme of the course may be to study the plants and animals of particular habitats, in which case it is relatively easy to move the emphasis towards why they are growing and living together. Ecological principles can thus be introduced in a way that is seen to be relevant to under-standing what the student is observing. Ecological concepts are, in this way, put into context by first-hand experience and are understood more readily.

At the end of such a course the participants are usually fascinated by ecology, and the skills of identification learnt along the way are a means to an end rather than an end in themselves. This was illustrated by a class taken to an urban woodland set between a coal mine and a large housing development. The object was to practise tree identification but one of the class noticed that most of the oaks were large and even-aged except for a few small saplings; the beech trees by contrast were either very old or comparatively young. Many reasons were discussed, giving the oppor-tunity of introducing new concepts (e.g. seedling mortality and causes, woodland structure, distribution of tree species with climate and soil, woodland types and classification) and detective work (e.g. looking for dead seedlings, cut stumps, examining the soil). The students decided that the most plausible hypothesis to explain these observations was past management. The question then arose as to how to test this hypothesis. Another visit to the wood was arranged and the group decided to measure the girth of all trees within an area, convert the girths to approximate age using forestry tables (Anon., 1986) and plot the resulting age frequencies (Table 4.2). From this it was argued that an original plantation of beech and oak was invaded by cohorts of seedlings associated with the two World Wars (age classes 41–50 and 71–80 years) probably as a result of opening of the canopy by felling for domestic and mining materials; removal of small oaks during the Second World War for coal mine props would account for the youngest trees being 71–80 years old. A search of historical records largely confirmed these findings and proved very rewarding for the participants.

To a large extent, there is nothing new in introducing ecology into natural history courses in this manner because it uses the well-tested teaching philosophy of adult education of exciting curiosity and encour-aging self-exploration of a problem.

A more recent development in adult education has been the expansion of educational 'holidays' or study tours. These are becoming increasingly popular, especially in place of the more traditional summer holiday spent on the beach. The true value of these study tours lies in the opportunities offered for longer-term exploration of a geographical area. Time can be used to investigate what is growing in an area before participants explore why plants and animals are observed in certain locations. (On single-day

Table 4.2. Frequency of different tree species in age classes (data collected February 1987 from Hemheath Woods, Staffordshire)

| Age classes (years) | Tree species | | | | |
| --- | --- | --- | --- | --- | --- |
| | Sycamore (Acer pseudoplatanus) | Beech (Fagus sylvatica) | Oak (Quercus spp.) | Ash (Fraxinus excelsior) | Others[1] |
| 0–10 | – | – | – | – | 7 |
| 11–20 | 4 | – | – | 1 | 7 |
| 21–30 | 29 | – | – | 3 | 6 |
| 31–40 | 24 | 5 | 1 | 11 | – |
| 41–50 | 7 | 2 | – | 6 | – |
| 51–60 | 4 | 6 | – | 7 | – |
| 61–70 | 1 | – | – | 3 | – |
| 71–80 | – | 1 | 3 | 1 | – |
| 81–90 | – | – | 2 | – | – |
| 91–100 | 1 | – | 4 | – | – |
| 101–110 | – | – | 9 | – | – |
| 111–120 | – | – | 2 | – | – |
| 120+ | – | 2 | 11 | – | – |

[1] Including Betula pubescens, B. pendula, Alnus glutinosa, Prunus avium, P. spinosa and Ilex aquifolium.

trips the desire to compile a comprehensive list of plants, a pass-time particularly enjoyed by the British, can consume a considerable proportion of the visit.) On a more mobile study tour, where new areas are visited each day, such exploration may be only verbal with ideas debated between participants; the success of this approach largely depends on the expertise of the leader to provide at least some definitive answers. On stationary trips run from a single base, it is possible to undertake simple ecological investigations which can call upon the resourcefulness of the most capable participant in designing simple tests using limited equipment. Courses are now being run in Britain (by, for example, the University of Surrey) where students are resident in an area for several weeks and, with little initial information, are helped to explore the location, and then to formulate and answer ecological questions. Similar approaches can be implemented from the laboratory with a group of adult students undertaking team research (Bagnall, 1975). In these cases, participants have full responsibility for their own learning, but, again, are answering questions arising from natural history observations.

## Practical habitat management

An alternative approach to the acquisition of knowledge via natural history is to involve people in practical management. In the courses organised at Keele University, participants are given responsibility for an area of the campus for which they create a management plan (or modify an existing one) and carry out necessary management. This is arranged as a series of Saturdays spread over eight months. Emphasis is on learning practical skills: tuition on the correct use of tools is provided, and experience is gained in management techniques such as tree felling and planting, and the creation of meadows and ponds. Inevitably this raises ecological questions, which can then be addressed by application of relevant theory. For example, several years ago a decision had to be taken as to how large to make a clearing within a dense woodland: this necessitated discussion of the species likely to invade various size clearings and the theory of island biogeography – immigration and extinction rates, and the number of species held by islands of different sizes with consideration as to whether woodland clearings could really be considered as 'islands'. In such cases the theories are not an end in themselves but are staging posts in trying to understand how the system would respond to management. Theory sessions are relevant in the field situation, acting as a catalyst aiding understanding.

## Conclusions

Ecology is an important component in adult education but, due to the
voluntary nature of the public's involvement, it has to be treated in a
specific way. Ecology is most appreciated when it is introduced into a line
of enquiry to help explain observed phenomena: all roads lead to ecology
rather than it being the starting point. Secondly, ecology has to be seen
to be relevant to a situation observed in the field rather than as an
academic exercise. With these points in mind, ecology has a bright and
important future in adult education and the life of the British public.

## Acknowledgements

My thanks to the adult students whose work contributed to the philosophy and examples quoted, and to Monica Hale for making this paper
possible. Financial assistance from the British Ecological Society and
PowerGen is gratefully acknowledged.

## References

Anon. (1986). *Practical Work in Farm Woods: 2 Woodland Survey and
    Assessment.* Agricultural Development and Advisory Service/Forestry
    Commission Leaflet P3018. Ministry of Agriculture, Fisheries and Food
    (UK) Publications.
Bagnall, R.G. (1975). Study-research groups as a method of teaching the
    natural sciences. *Journal of the International Congress of University
    Adult Education*, **XIV**(2), 21–41.
Bland, D. (1981). A survey of extramural students in the Liverpool area. *The
    Tutors' Bulletin for Adult Education*, **4**(1), 14–18.
Feltwell, J. (1983). Awareness of wildlife conservation among adult education
    students. *The Tutors' Bulletin for Adult Education*, **6**(1), 8–11.
Gronemeyer, M. (1987). Ecological education a failing practice? or: Is the
    ecological movement an educational movement? In ed. W. Leirman and
    J. Kulich, pp. 70–83. *Adult Education and the Challenge of the 1990s*,
    London: Croom Helm.
Jones, B. (1986). The teaching of controversial issues in adult education.
    *Studies in the Education of Adults*, **18**(1), 3–10.
Tansley, A. (1939). *The British Islands and their Vegetation*. Cambridge:
    Cambridge University Press.
Tansley, A. (1949). *Britain's Green Mantle*. London: Allen and Unwin.
Universities Council of Adult Education (1972). University Adult Education in
    the later twentieth century: the UK. *Journal of the International Congress
    of University Adult Education*, **XI**, 88–101.
Wiltshire, H. (1983). The role of the University Adult Education Department.
    *Studies in Adult Education*, **15**, 3–10.

# 5

# New opportunities for ecology education in the United States

ALAN R. BERKOWITZ

*Head of Education, Institute of Ecosystem Studies and Education Section Chair 1990, Ecological Society of America*

## Introduction

This paper is primarily directed to professional ecologists interested in ecology education at the primary and secondary levels. My purpose is to explore how ecologists in the United States might contribute to, guide and help improve the quantity and quality of education in ecological science in primary and secondary schools. What can ecologists, such as the membership of Ecological Society of America (ESA), do? In order to answer this question, we must address several related questions: 1. How is ecology being taught at the current time? 2. What do we mean by excellence in ecology education, i.e., what are we striving for? 3. What are some promising new directions we can take in developing programs that address these goals? 4. Where should we be exerting our influence in advocating and working for change?

Work towards excellence in ecology education must go beyond the concerns of professional ecologists. While ecologists are concerned with the apparent lack of coverage of our discipline and the lack of accuracy when it is covered, many issues and challenges face our nation's education system as a whole. We must be sensitive to and understand the impediments and challenges to improving science and environmental education if we are to be effective players in this theater. Likewise, we must identify and build upon the successes and strengths of the education community at large, and of the science and environmental education arenas in particular. In so doing, new ecology education initiatives will make significant contributions to the attainment of the pressing goals in these broader fields as well.

Ecology is considered in this paper as the scientific study of organisms as they interact with each other and the physical environment, as distinguished from the broader field of *environmental studies*, which includes all the social and scientific disciplines pertaining to the environment. Thus, ecology education is distinct but clearly related to other educational arenas. The goal of ecology education is to foster ecological literacy, defined as: 1. an understanding of the scientific process as applied in ecology; 2. a familiarity with the ecological processes at work in one's local environment; and 3. sufficient familiarity with ecological principles to be able to understand the basic ecology of environmental problems in other regions. This goal contrasts to that of environmental education, which fosters an informed, concerned and environmentally responsible public. The tremendous upsurge in concern about the environment, and about faltering scientific literacy and its implications for responsible action and international competitiveness, encourages redoubled efforts to enhance ecology education as a part of broader education reform.

## Current status of ecology education in the United States

National leadership in education in the United States is diffuse, with most of the key decisions about curriculum and instruction being made at the local level. However, the curriculum adoption procedures of certain crucial states such as Texas and California has a profound influence on education nationwide. In these states, books or programs are selected to be used throughout the commonwealth. Publishers mold their texts and other products in order to win approval in these states. This, in turn, constrains the nature of materials available elsewhere.

The United States does not have a national curriculum in any area. Each state is free to mandate (or not) the curriculum of the schools within that state. Most curriculum in the US is set at the local level. Thus, there is *no* national policy or curriculum for ecology or environmental education. The Federal Government has had some involvement in environmental education in the past, but until recently this has been quite minimal. The recently enacted National Environmental Education Act (November 1991) establishes an Office of Environmental Education in the US Environmental Protection Agency and several other programs, internships and grants for environmental education. The new grant programs initiated by this act provide a level and continuity of funding not previously available to environmental education projects.

Another form of leadership in education is provided by several agencies of the Federal Government through their support of innovative education programs. Many of these programs have strong ecological and/or environmental components. For example, in the mid 1980s the National

Science Foundation's (NSF) Informal Science Education Program funded 27 projects of which eight included ecology (NSF, 1987a). In the NSF Materials Development Program, six of the 19 projects at the elementary level and seven of 33 projects at the secondary level included ecology (NSF, 1987b). Similarly, the US Department of Education's National Diffusion Network listed 12 programs, six of which have some relation to ecology or environmental education (USDE, 1987).

State governments have a great deal of influence on the teaching of ecology. Though no surveys of the status of ecology education are available, several regarding environmental education reveal the extremely variable role states play across the nation (Table 5.1). By the year 1987, only 19 of the 50 states had significant legislation and/or policies regarding environmental education (Marshall, 1987). In 17 states, some form of environmental education had been legislated as mandatory, but only five required that teachers take courses in environmental education in order to be certified. These numbers represent a marked increase over the past decade; current trends suggest that many states will strengthen their environmental education programs, mandates and requirements in the coming years.

Several organizations have made significant contributions to environmental education at the national level. Perhaps most visible are the curriculum supplements Project WILD and Project Learning Tree. Project WILD is a joint endeavor of the Western Association of Fish and Wildlife Agencies and the Western Regional Environmental Education Council. Since its inception in 1983 over 240 000 educators in 49 states and two foreign countries have attended Project WILD workshops and an estimated 20 000 000 students have participated (Charles, 1990). In response to a 1990 survey of teachers, 97.8% reported improved student awareness, knowledge, skills and/or attitudes about the environment. Over

Table 5.1. *Adoption of environmental education by states*

| Policy or feature | Number of states (of 50) |
|---|---|
| Legislative mandates | 17 |
| Learning standards specified | 16 |
| Teacher certification requirements | 5 |
| State environmental education specialists | 17 |
| State definition of environmental education | 19 |

Based on a National Wildlife Federation/National Audubon Society survey (Marshall, 1987).

25% of the teachers polled reported that Project WILD was their *only* source of environmental education materials.

Project Learning Tree is a joint project of the Western Regional Environmental Education Council and the American Forest Foundation. Since it was first published in 1975 over 200 000 educators in 49 states, five Canadian provinces and three foreign countries have been trained and an estimated 10 000 000 students have participated (Project Learning Tree staff, pers. commun.).

Both Project Learning Tree and Project WILD have made tremendous contributions to environmental awareness and, to a certain extent, to an understanding of basic ecology among school children and their teachers. However, neither was intended to be, nor serves as, a rigorous and comprehensive ecology curriculum.

There are a number of national organizations advocating improvement and reform of science education. Two are particularly noteworthy with regard to ecology education. The American Association for the Advancement of Science (AAAS) is sponsoring Project 2061. The first stage of this ambitious project resulted in a 100-page publication, *Science for All Americans*. Here, teams of scientists tried to lay out what every high school student should know about science. Ecology is well represented in the publication. In the second stage, now underway in Philadelphia, Pennsylvania; McFarland, Wisconsin; rural Georgia; San Antonio, Texas; San Francisco, California; and San Diego, California, teachers are developing model curricula to teach the concepts delineated in stage one. In stage three, beginning in 1993, school systems are expected to start implementing the new programs. This is certainly one of the most long-range and thorough of the curriculum reform projects currently on the national scene, and quite worthy of participation from ecologists.

The National Science Teachers Association's (NSTA) Scope, Sequence and Coordination (SSC) project is another example. As the principal professional society of science teachers in the country, NSTA has undertaken a very aggressive program to restructure science teaching in grades 7–12. The emphasis is on cross-disciplinary approaches to understanding real problems. Each year, students learn in all science subject areas, with work progressing from concrete to abstract at the higher grades. The project is being developed at sites in California, Iowa and Texas, with Puerto Rico and North Carolina scheduled to join in soon. There is much ecology being incorporated; for example, the program in Davenport, Iowa, is a science–technology–society approach to environmental problems.

A number of professional environmental education organizations such as the North American Association of Environmental Education and the newly revitalized Alliance for Environmental Education are attempting to provide leadership in this area. However, neither group has a strong representation of ecologists in their constituency or involved in their programmatic efforts. As a result, the science of ecology often is omitted or is promoted and taught by non-ecologists.

The Ecological Society of America (ESA), the principal professional society of ecologists in the United States, may be in a position to provide leadership in ecology education nationally. In 1988 the ESA formed an Education Section with a current membership of 375. Over 1050 of the Society's 6800 members responded to a recent survey about education, with 646 ecologists volunteering to get actively involved in pre-college education activities (Berkowitz, 1988). The Education Section has embarked on a number of activities and projects (Box 5.1). It is our hope that through these diverse efforts we will help secure a solid place for ecology in the educational system of the future.

---

Box 5.1. Projects of the Education Section of the Ecological Society of America

1. **Promoting ecology education**

    Advocating improved federal and state policies, programs and mandates.

    Encouraging universities to recognize the educational efforts of ecologists.

    Providing awards for excellence in ecology education.

2. **Publications**

    *Section Newsletter*, giving current information and networking.

    *Ecology Education for Children: A Handbook for Ecologists*, an information brochure for professional ecologists (draft is available, final version in preparation).

    *Ecology Education for Children: A Framework for Excellence*, synthesizing the section's recommendation for teaching ecology at the primary and secondary level (draft is available, final version in preparation).

    *Ecology Education for Children: Strategies for Teacher Education*, recommendations and strategies for enhancing the ecology education of our teachers (in preparation).

*Ecology Education for Children: Concepts for Curriculum*, a recommended framework for curriculum content for primary and secondary schools (under consideration).

*Directory of Ecologists for Education*, a cross-referenced list of ecologists willing to get involved in education (available on computer disk).

*Experiments in Ecology*, a compendium of tested experiments for teaching ecology.

*Careers in Ecology*, a guide to jobs in ecology.

3. **Committees**
   *Teacher Enhancement and Training*
   *Undergraduate Education*
   *Secondary Education*
   *Education of Young Children*
   *Minorities in Ecology*

4. **Symposia, workshops and special programs**
   Programs at the national meetings of the society, for example 'Ecology Education for Undergraduates: Current Perspectives and Future Directions' (symposium), 'Ecology Education in Primary and Secondary Schools – Towards an Agenda for Excellence' (symposium), 'Experiments in Ecology – Successful Laboratory Exercises' (workshop), and a model ecology camp.

5. **Affiliations**
   National Science Teachers Association (NSTA)
   National Association of Biology Teachers (NABT)
   North American Association of Environmental Educators (NAAEE)
   Alliance for Environmental Education (AEE)

**Excellence in ecology education**

The Ecological Society of America is attempting to define what we mean by excellence in ecology education. Our current ideas, spelled out in the draft document *Ecology for Children: A Framework for Excellence*, include the following principles. Our educational system should provide students the opportunity to:

1. have repeated, hands-on experiences with organisms in their environment;

2. conduct ecological research;
3. learn the relevance of ecology to human concerns;
4. study ecology in their local environments;
5. learn accurate and contemporary ecological principles and concepts;
6. learn irrespective of their learning styles, backgrounds and dispositions towards science;
7. learn about the work and life of ecologists from real scientist role models;
8. take an interdisciplinary approach to understanding the environment;
9. learn the relationship between local, regional and global scales; and
10. be helped to transfer skills and dispositions learned in ecology education to other disciplines and to their everyday lives.

While addressing these goals it will help us to remain aware of the current issues that define the cutting edge for other areas of education in the future. From the broad-scale movement for reform within the education community at large we see consistent trends for: 1. emphasizing depth over breadth, 2. empowering teachers, 3. promoting interdisciplinary study, 4. providing opportunities for hands-on learning, 5. addressing student pre-conceptions, 6. serving the full spectrum of learners, and 7. involving professional scientists *and* educators. Excellence in ecology education and in education in general should go hand-in-hand.

The pressing concerns facing the environmental education arena should also remain clear in our minds: fostering an appreciation of nature while empowering citizens for stewardship. Environmental educators consider a full range of approaches to this very broad objective, including sensory awareness, physical challenge experiences, and the use of the arts and literature in addition to scientific discovery. This is in clear contrast to the focus of ecology education on *scientific* understanding. Likewise, in addressing the stewardship goal, attention is paid to the social, political and moral contexts of the science of ecology. The role ecologists and ecology education play in this arena will depend on our willingness to build these intellectual and collaborative bridges.

## New programs for ecology education

This section will describe some of the most exciting themes and practices in science and environmental education being undertaken currently to show how ecology education programs might be developed to address the goals and concerns outlined above. Each program mentioned excels in attaining several of the goals for excellence in ecology education. The name and address of a contact person for each program highlighted is given at the end of this chapter.

### Denver Audubon Society's Urban Ecology Project

One of environmental education's greatest strengths is in its capturing of students' curiosity through immersion in their immediate environment. The material of study is right underfoot, overhead and all around you. Children must learn that ecology is a science that can be applied to understanding any environment – local ecosystems should be emphasized wherever possible. A sterling example of a program that gives children repeated, hands-on exposure to their immediate environment is Denver Audubon Society's Urban Ecology Project. This program uses hands-on outdoor activities from Outdoor Biology Instructional Strategies (OBIS) to teach 4th or 5th graders (8–10 year olds) about ecology. Volunteers run four one hour programs per year in environments within the students' urban environment. In 1990, 840 volunteers taught 8600 students in eight cities through this program.

### Cornell University's Environmental Sciences Interns Program

Minorities and women are underrepresented in all of the technical, mathematical and scientific fields. Programs targeted directly at these groups are essential, and evidence suggests that they work. For example, Cornell University offers an Environmental Sciences Interns Program in which fifteen high-ability high school students from areas with large minority populations participate in a six-week residential program. The program focuses on research, careers, community outreach and ethics. Follow-up activities in schools and 4H clubs engage the participants in outreach with children and teachers in their home communities (4H is a Cooperative Extension supported youth program in agriculture, home economics, and conservation). This program succeeds with its combination of direct experience with research and the environment, exposure to real-life scientist role models, and extension to the students' social and personal context in their life at home.

### Science by Mail

An important need in promoting equity in education is to change children's stereotype of a scientist as a mad caucasian male in a white coat.

They must learn that scientists are a diverse group of people with diverse human concerns, feelings and backgrounds. Science by Mail, a project of the Museum of Science in Boston, Massachusetts, matches groups of one to four children with a volunteer mentor scientist. The groups receive three challenge packages and then correspond with their mentor scientist in solving them. In 1989, 3000 groups in 13 chapters corresponded with 850 scientists in this exciting program. Ecology is not a prominent theme in the Science by Mail program; however, the ESA could use this as a model program to provide professional ecologist role models to get away from the 'long beard and sandals' caricature of the 'typical ecologist'.

## National Geographic Kids Network

Ecology, when applied to environmental problems, involves a broad range of scales from local to global. Likewise, it necessitates an interdisciplinary approach to problem solving. The magnitude of global problems poses both an enticement and a challenge to ecology education. New telecommunications technologies allow creative approaches to dealing with such large-scale phenomena. The National Geographic Kids Network is a joint effort of the National Geographic Society and the Technical Education Research Centers. Through this program, students share data via a telecommunications network to examine large-scale problems such as: acid rain, weather, and water quality. Unit scientists communicate directly with students throughout the network. This program is especially exciting in that it combines several of the elements of excellence: scientist role models, relevance, interdisciplinary approach and multiple scales.

## Science Curriculum Improvement Study (SCIS)

Children need to experience knowledge-building in ecology as in all science learning. It is particularly difficult to devise experimental systems for teaching ecology, since many of the systems are complex, responses are subtle and results take a long time to appear. Furthermore, there is great misunderstanding about the nature and practice of science and scientists. Ecology is no exception. Students need experiences doing, not just reading about, science. This is not a new realization and certainly is not unique to ecology. Numerous studies over the past three decades have documented this fact. An excellent example at the elementary level is the Science Curriculum Improvement Study (SCIS). Included is an entire workbook on ecosystems, replete with interesting and well articulated studies. Unfortunately, the use of such programs requires more material, training and logistical support for elementary teachers than is available in most instances, severely limiting the extent of their utilization. However, new efforts drawing upon the strength of such programs hold much promise.

### Biological Science Curriculum Study (BSCS)

The complexity of ecological concepts makes the discipline quite prone to misconceptions. This problem is made worse by the apparent familiarity of many of the objects of study, e.g., ponds, animals, communities. Children bring many pre-conceptions, some based on anthropomorphisms and others from romanticized notions about nature, to this subject. We must help students examine and revise their conceptual frameworks about the world. A very exciting approach to building such considerations into an ecology-related curriculum is the BSCS's 1991 initiative for elementary schools, *Science for Life and Living*. The essential elements of their approach include: explicit consideration of students' pre-conceptions; emphasis on overarching conceptual frameworks and unifying ideas; and study of fewer concepts over longer time periods in greater depth.

### Institute of Ecosystem Studies' Eco-Inquiry Curriculum

The Institute of Ecosystem Studies has developed the Eco-Inquiry Curriculum for upper elementary students. Eco-Inquiry includes five activity clusters designed to promote active and reflective learning throughout a semester. The content of the curriculum focuses on how materials cycle and energy flows among the components of an ecosystem. Eco-Inquiry does not, however, focus solely on ecology content. What makes the program unique is that it emphasizes how the world of science itself functions. Eco-Inquiry transforms the classroom into a research community. Throughout Eco-Inquiry scientific habits of mind are targeted such as controlling impulsivity, seeking empirical evidence to support claims, and the willingness to change opinions in light of new data. Activities help students understand what thinking dispositions are, and internalize a sense of their importance and the criteria for success. It is not expected that students will master all of these sophisticated dispositions during one term in elementary school. However, by communicating clearly the role of dispositions in scientific work *and* in everyday decision making, and by encouraging students to be aware of and improve their own dispositions, the curriculum aims to foster growth toward adoption and transfer of these crucial affective traits.

## Working for change

Ecologists dedicated to improving ecology education in the United States can work at any level of the system with significant effect. Below are brief descriptions of those activities that might have most impact in promoting excellence in ecology education in the future. The ESA is just beginning to explore these avenues for its future efforts.

### National environmental education initiatives

Environmental education often is 'sold' as including ecology. There is a great increase in attention being paid to environmental education in the US, including the new Office of Environmental Education in the EPA, new monies for programs, and consideration of new mandates and standards (e.g., the Alliance for Environmental Education's Standards in Environmental Education Project). Ecologists must get involved in order to assure that ecology is well represented in any new initiatives that are forthcoming.

### State curriculum and textbook adoption

Each state can mandate the curriculum of its schools. Thus, working with the state education department can be a crucial way for ecologists to exert influence. Attention should be paid to mandated learning outcomes and curriculum content, standardized tests, and instructional syllabuses.

### State teacher education requirements

Ecologists should work with state legislatures and education departments to strengthen the training of pre-college teachers in the sciences. All teachers should be required to take a minimum number of credit hours in environmentally oriented science courses in gaining their teaching credential. Likewise, school systems should provide in-service training for teachers in the science behind environmental problems. Environmental educators also should be required to take courses in ecology as part of their professional training. States might consider establishing an environmental educator credential in order to assure adequate training.

### Creating new programs and curricula

Ecologists willing to put in the time and effort will find it enormously rewarding to work in a curriculum development team with professional educators. Professional ecologists are constrained in pursuing such activities by pressures from their department to pursue strictly research interests. This is a significant hurdle that must be overcome if we are to be full partners in crafting the curricula of the future. Indeed, as in any discipline, some of the most exciting developments in ecology education are likely to spring from such collaborations.

### Conclusions

New developments in ecology education in the United States are filled with promise. The current level of interest in the environment, the high level of commitment to science and environmental education at the local and national level, and the upsurge of activity among professional ecologists all bode well. We are not likely to have a national curriculum in any subject in the near future, nor are all states likely to mandate rigorous ecology education. However, the prospects for excellent programs, curricula and texts are bright. Part of this is attributable to the very decentralization that makes national leadership so difficult to discern or implement. While professional ecologists can and should strive to promote ecology education at the state and national level, educational excellence ultimately is achieved in classrooms, on nature walks, in school yards and at home. People developing and promoting the new generation of ecology education programs, some of which are highlighted here, work primarily at this level as well. This assures that their products will be useful and useable, tailored to the local ecological, sociological and institutional conditions, and thus more likely to succeed.

### Acknowledgements

On behalf of the Ecological Society of America (ESA) and the Institute of Ecosystem Studies, I would like to extend my gratitude to the symposium organizers, the INTECOL meeting organizers, and our gracious

Japanese hosts. Thanks also to Kathleen Hogan for her invaluable comments.

## References

Berkowitz, AR (1988). Ecological Society of America, Education Section Survey. Millbrook, NY. (unpublished).

Charles, C. (1990). *Project WILD: Report of Program Activities from a National Perspective*. Boulder, CO.

Marshall, K. (1987). State legislation for environmental education. Washington, DC: National Wildlife Federation and National Audubon Society. (unpublished).

NSF (1987a). *Summary of Grants, FY 1984–86: Informal Science Education Program*. Washington, DC: National Science Foundation. Directorate for Science and Engineering Education.

NSF (1987b). *Summary of Grants, FY 1984–86: Instructional Materials Development Program*. Washington, DC: National Science Foundation. Directorate for Science and Engineering Education.

USDE (1987). *Science Education Programs That Work*. Washington, DC: US Department of Education, Office of Educational Research and Improvement.

## Program and project contacts

**Alliance for Environmental Education (AEE) – Environmental Education Standards Project**
Charles Roth
EDC, Inc.
55 Chapel St.
Newton, MA 02160
(617) 969-7100

**American Association for the Advancement of Science (AAAS) – Project 2061**
Dr. F. James Rutherford
Chief Education Officer and Director, Project 2061
AAAS
1333 H St., NW
Washington, DC 20005
(202) 326-6620

**Biological Science Curriculum Study (BSCS)**
Dr. Rodger W. Bybee
Biological Sciences Curriculum Study
830 North Tejon St., Suite 405
Boulder, CO 80903
(719) 578-1136

**Cornell University – Environmental Sciences Interns Program**
Dr. Marianne Krasny
Project Director
Cornell Environmental Sciences Interns Program
Department of Natural Resources
Fernow Hall
Cornell University
Ithaca, NY 14853
(607) 255-2814

**Denver Audubon Society – Urban Ecology Project**
Karen S. Hollweg
Project Director, Urban Ecology Program
Denver Audubon Society
3000 South Clayton, #207
Denver, CO 80210
(303) 757-7858

**Institute of Ecosystem Studies – Eco-Inquiry Curriculum**
Kathleen Hogan
Eco-Inquiry Project Director
Institute of Ecosystem Studies
Millbrook, NY 12545
(914) 677-5358

**National Geographic – Kids Network**
Dr. Candace Julyan
Director, National Geographic Kids Network Project
Technical Education Research Centers (TERC)
2067 Massachusetts Ave.
Cambridge, MA 02140
(617) 547-0430

**National Science Teachers Association (NSTA) – Scope, Sequence and Coordination Project**
Dr. Robert Yager
SSC Program Director
NSTA
1742 Connecticut Ave, NW
Washington, DC 20009
(202) 328-5800

**Project Learning Tree**
Kathy McGlauflin
Director, Project Learning Tree
1250 Connecticut Ave. NW
Suite 320
Washington, DC 20036
(202) 463-2468

**Project WILD**
Dr. Cheryl Charles
Director, Project WILD
Post Office Box 18060
Boulder, CO 80308-8060
(303) 444-2390

**Science by Mail**
Stephen Brand
Coordinator of Science By Mail International
Museum of Science
Science Park
Boston, MA 02114-1099
(800) 729-3300

**Science Curriculum Improvement Study (SCIS) and Outdoor Biology Instructional Strategies (OBIS)** (used in Denver Audubon's programs) available from:
Delta Education
Box M
Nashua, NH 03061
(800) 258-1302

# 6

# Ecology education and field studies: historical trends and some present-day influences in Britain

S.M. TILLING

*Field Studies Council, Montford Bridge, Shrewsbury SY4 1HW, UK*

## Introduction

Ecology is a difficult discipline to teach. It places considerable demands on the education system – on researchers, teachers and administrators alike. This is particularly true if field studies are regarded as an indispensible part of an ecology course. It is likely, therefore, that ecology is not being taught effectively in a significant number of secondary schools in Britain. This may present colleges and universities with a problem in the future; they will be expected to service the growing demands of a more environmentally conscious population. This chapter examines the historical and present-day problems associated with ecology, both in research and teaching, and then summarises the likely state of contemporary ecology teaching in Britain. Lastly, some suggestions are made towards improving the state and profile of ecology in the British education system.

## The problems

### The perception that ecology is a 'soft' science

In the early years of this century ecology as a scientific discipline in Britain was almost inseparable from 'recreational' natural history (Berry, 1987). If the discipline could be summarised as looking at 'why is what where' in the natural world (Berry, 1989), much of the early work concentrated on mapping what is where. Britain was fortunate in that it had numerous competent natural historians and flourishing local natural history clubs, many of which became the backbone for these early initiatives. They were later to form the national societies such as the British Ecological Society (Sheail, 1987). However, much of this early work was routine, involving extensive and repetitive vegetation mapping, and some lacked the rigour expected of a scientific discipline. Tansley (1951), one of the architects of modern British ecology, remarked that 'a good many of the papers published during the early part of the century were ... rather trivial and some of them decidedly slovenly. Besides stimulating many good biological minds, ecology had a great attraction for weaker students, because it was so easy to describe particular bits of vegetation in a super-ficial way, tending to bring the subject into disrepute'. The study of natural history, and, therefore, ecology in its early years, was also the domain of the relatively privileged. Francis Buchanan-White, another of the early British ecologists, was typical in 'neither requiring nor caring to practise his chosen profession as a medical man' (Sheail, 1987). Though the social restrictions placed on ecology in its early days were not peculiar to that subject – mass education in Britain is relatively recent, only becoming the norm after the introduction of the Education Act of 1944 – this, and the 'amateur' nature of much of its practice, meant that ecology was easily characterised as a lightweight subject, akin to stamp collecting. For example, Tansley's colleagues would claim that ecology 'is only the old natural history masquerading under a high-sounding name – and not always good natural history at that!' (Tansley, 1951). This impression has persisted. For example, one American scientist recently derided the 'fascination with birds and gardens, butterflies and snails which was so characteristic of the prewar upper middle class from which so many British scientists came'.

## Reductionism and determinism: the twin towers of science

Ecology also presents scientists with a number of methodological and conceptual problems, some of which are very difficult to resolve. In particular, 'good' science has always been subject to reductionism and determinism. Inevitably, research is easier to carry out if the subject under scrutiny can be considered in isolation and the results, interpretation and knowledge can be slotted subsequently into an existing framework. Scientists tend to adopt the law of parsimony (Occam's razor), to choose the simplest possible hypothesis even though others are possible. However, ecology does not lend itself readily to simplification or compartmentalisation. Its conceptual jigsaw is many-faceted and multi-dimensional. This has not prevented ecological scientists from striving for unifying principles and laws, largely in an attempt to put some sort of order into a seemingly endless list of unrelated facts and figures (e.g. Southwood, 1977, 1988; Berry, 1989). The breadth of knowledge that a competent ecologist is expected to acquire is formidable; in a recent survey British Ecological Society members listed well over 50 concepts that were important in ecology (Cherrett, 1989). This is an inevitable consequence of dealing with the natural world. Pantin (1968) claimed that 'physics and chemistry have been able to become exact and mature because so much of the wealth of natural phenomena is excluded from their study'. Furthermore, ecologists are often expected to become familiar with a range of subjects that are not associated with a formal biology education; aspects of physics, meteorology, engineering, chemistry, geomorphology, hydrology, archaeology and history are consulted frequently and invoked to provide an adequate interpretation of patterns and trends revealed by ecological research. Inevitably, the numerous skills and techniques (formal and improvised) employed in ecological research, both in the laboratory and the field, are also a source of confusion as are the mathematical and statistical approaches used in presentation and analysis.

## Ecology as a foreign language

The growth of ecology as a 'new' science and its interspersion with other disciplines has led to a concomitant proliferation of novel terms, definitions and descriptions. Although this is a problem facing any inexperienced student of sciences it is difficult to think of any other discipline that involves familiarisation with such an array of terminology.

## Fieldwork: the Achilles heel of ecology?

Ecology demanded a radical change in approach by biological scientists. It represented a break away from the entrenchment in the laboratory, museum and herbarium in which nearly all 'serious' biology had been practised previously. It provided a link between the formal disciplines such as physiology and morphology and the natural environment in which the plants and animals existed.

Recently, there has been a retreat back into the confines of the research laboratory, away from the field-based natural history foundations of the subject. This is not surprising. The immense body of knowledge, and the need to apply scientific rigour to ecological research, has led to increasing fragmentation and specialisation within the subject. It is, however, a trend that would have dismayed the early founders of the subject, most of whom based their research on a wide-ranging knowledge of field-based natural history. Tansley (1951), when considering the characteristics of a 'good ecologist', thought that he should have an elementary knowledge of many branches of biology (and other non-biological subjects), but 'above all ... must have a practical experience of fieldwork – the more the better. Ecology is essentially a 'field subject': progress cannot be made in it without constant observation (and frequently experiment) in the field though some of the ecological problems met with in the field have to be attacked in the experimental garden or the laboratory, or both'. Any move back into the security of the laboratory may be particularly damaging for the long-term development of ecology; one of the few dictums that can be applied in ecology is that most first-rate ecologists are also first-rate field naturalists.

**Where does this leave the educationalists?**

The complexity of ecology presents educational administrators and the teaching profession in secondary education with an almost impossible task. The Standing Conference on University Entrance (1977) looks for the following in an ideal candidate going on to higher education:

students must have acquired some feeling for the overall classification and morphological diversity of living organisms, whenever possible relating them to the environment in which they are living ... It follows from this that there ought to be an appropriate amount of study in the field. Such study should lead towards an understanding of ecological principles, in particular the concept of the ecosystem and of the ecological niche, food chains and webs, dynamic nature of plant and animal populations and energy flow through ecosystems.

Therefore, the ideal 'ecologist' emerging from the British secondary education system should be a competent field-based natural historian with a good knowledge of ecological 'principles' and, at the very least, a basic competence in one or more complementary subjects such as physics, chemistry, meteorology, history and statistics. How else could the feeling, understanding and ability to interpret be achieved? How, therefore, should we teach 'ecology' within the confines of the school and a formal teaching timetable?

Clearly, it is impossible to achieve the ideal. A degree of compromise is necessary. But that does not preclude the need to search for the best alternative – one that optimises the use of available resources. Booth (1979) listed five factors that prevented the teaching of ecology in schools: the nature of the subject; examination syllabuses and papers; lack of facilities; difficulties and confusion about how to teach ecology; and lack of confidence of teachers. Are these still important in present-day ecology education?

### The nature of the subject

The complexity of ecology has already been described. Unlike other branches of biology, such as genetics and physiology, ecology is not condensed easily into a few underlying laws or principles, and cannot be considered (efficiently) in isolation from the rest of biology and other complementary subjects. Ecology will never be an easy, or a 'convenient', subject to teach – despite its reputation as a 'soft' science.

### Curriculum developments, examination syllabuses and papers

Although ecology was dropped (briefly) from the compulsory core of the A level syllabus of one examination board,* it remains as a separate topic in most. This has been criticised repeatedly (e.g. Booth, 1979; Harper, 1982). Inevitably, attempts to teach ecology in isolation produce a framework that lacks balance and cohesion, and one that will lead to confusion in the pupils. Unfortunately, the tendency to regard ecology as a separate subject, within biology or environmental science, still persists in many syllabuses. This is also true for the new science National Curriculum, which was introduced for the first time in British schools in 1989; here, the 'ecology' component was contained almost exclusively in two of the 17 science Attainment Targets originally laid out by the Department of Education and Science[†] (DES, 1989b; see also Hale and Hardie, this volume). Ecology still appears as an extraneous activity in later versions of the science curriculum. The teaching of ecology in isolation also makes it more likely that courses can be developed in which no ecology is taught; unfortunately, this option already exists in one of the two models open to 14–16 year olds in the new science National Curriculum in Britain. The need for fieldwork in ecology teaching has already been alluded to and, therefore, it is important that the status of out-of-classroom activities is given high priority. Surprisingly, although fieldwork is often recommended as a useful adjunct to a biology course (at A level, at least) it is rarely a compulsory requirement, as it is in many geography syllabuses.

---

* A levels. Upper secondary school syllabuses undertaken by British post-16 students (17–18 years old); preceding entry to higher education.
† Department of Education and Science. The government department which has overall responsibility for all education in British schools and colleges.

This clearly has an effect on the amount of biology fieldwork undertaken by schools, particularly work carried out away from the vicinity of schools (Figure 6.1a, b). There are some encouraging developments, however. The increasing importance given to environmental education in Britain and Europe, as highlighted by recent government documents (e.g. Department of Education and Science, 1989a) has raised the profile of ecology and fieldwork in our schools.

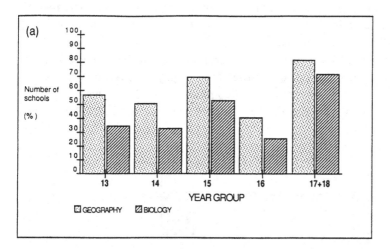

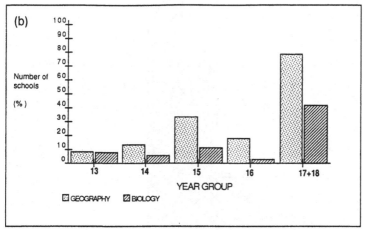

Figure 6.1. (a) Proportion of secondary schools that were carrying out geography and biology fieldwork (local and out of county (i.e. away from the school)) with different year groups in a survey of schools in Buckinghamshire, England. (b) As above, but only including survey results for out-of-county fieldwork. Graphs adapted from Arthern (1985).

*Facilities*

There has been a substantial increase in field studies facilities in Britain since the early 1950s. At present, there are over 400 centres and numerous organisations in Britain that offer 'environmental education' facilities and supporting resources; most education authorities and schools have access to local and out-of-county field centres (Figures 6.2 and 6.3; Herbert *et al.*, 1972; Environment Council, 1990). In addition, the opportunities for local fieldwork, particularly in urban areas, have been increased by a growth in wildlife trusts (many of which now have education officers) and other educational schemes.

There is growing evidence, however, that many such schemes are under threat, or have disappeared altogether, largely as a result of the withdrawal of central government and commercial funding. This trend is the subject of surveys in Britain, but evidence from the mid-1970s shows what

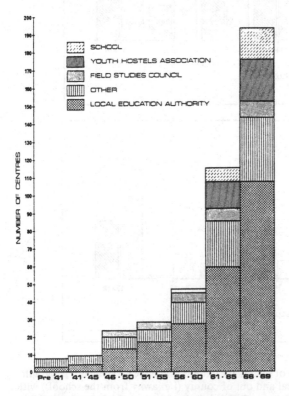

Figure 6.2. Numbers of centres of different types in use for field studies in successive periods up to 1969. Taken from Herbert *et al.* (1972). Reproduced with the permission of the Field Studies Council.

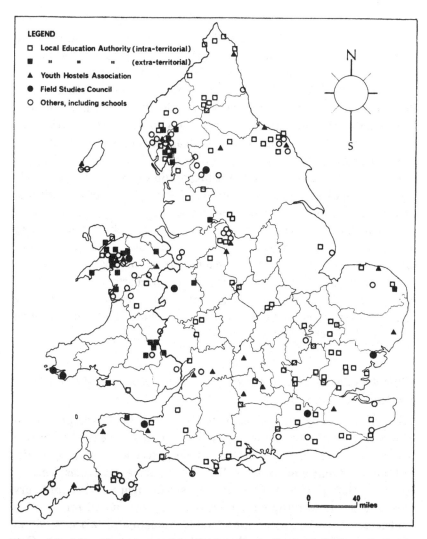

Figure 6.3. Map of centres used for field studies in England, Wales and the Isle of Man in 1970. Taken from Herbert *et al.* (1972). Reproduced with the permission of the Field Studies Council.

can happen if local authority or central government funding is withdrawn. During this period, in which there were reductions in financial support for field excursions, there was a significant decline in the number of students attending field courses in Field Studies Council centres*

* Field Studies Council. An environmental education charity which administers ten field centres in England and Wales; students from schools and colleges use these centres for ecology and geography field courses.

(Crothers, 1987). A more recent survey of Local Education Authorities in England and Wales shows that there is a continuing threat to staff, training, facilities and resources, largely caused by recent legislation, which has devolved financial responsibility for funding such activities away from Local Authorities, into the schools themselves (CEE 1992). Priorities are now set at school level. Anecdotal evidence pointing to a decline in facilities and staff supporting fieldwork activities is provided by falling memberships in organisations such as the National Association of Field Study Officers (J.E.B. Bebbington, pers. comm.).

The amount of time available for preparation and provision of field-work also appears to be a major disincentive in schools. In fact this was cited as the main problem by teachers in one of the few systematic surveys of fieldwork in secondary schools in Britain, with lack of finance being the next most important factor (Fido and Gayford, 1982). Health and safety considerations are likely to be more prominent now; strict govern-ment guidelines have been introduced recently (Department of Education and Science, 1989c) and these, plus the cautionary advice given by teachers' trade unions on taking pupils out of school, have had a negative effect on the level of fieldwork provision.

### *Difficulties and confusion about how to teach ecology and lack of confidence of teachers*

The association between ecology teaching and fieldwork presents a considerable hurdle for those seeking to establish a more prominent role for ecology in school education. Most biology teachers are not specialist ecologists and only a minority are likely to be natural historians. This led Harper (1982) to argue that 'employers have no right to expect ordinary biology teachers to possess the exceptional skills required for teaching field ecology'. There may be considerable sympathy for this view within the teaching profession. A number of solutions have been proposed. One of the most radical of these, and possibly the most threatening to ecology, is to exclude the (already optional) fieldwork component from biology teaching (King, 1989). Most of the evidence for or against fieldwork is anecdotal but some research does suggest that the majority of biology A level teachers in secondary schools, until quite recently at least, attributed some importance to fieldwork (Figure 6.4). This is significant because other research has indicated that teachers' attitudes were the most signi-ficant factor in determining the quantity of fieldwork carried out, out-weighing other problems such as time, facilities, discipline, etc. (Fido and

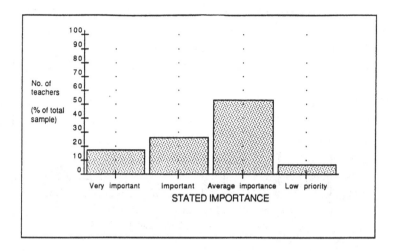

Figure 6.4. Proportion of biology teachers (percentage of total sample) attributing different levels of importance to fieldwork. Adapted from Gayford (1985).

Gayford, 1982). Less reassuring was the amount of time spent on fieldwork; between 50% and 75% of biology teachers (depending on the age groups being taught, but including A level) spent less than 20 hours field teaching each year (Figure 6.5). Teachers' attitudes also affect the quality of task setting. Pupils of teachers who attributed least importance to fieldwork spent more time on practising basic skills and learning the factual and theoretical component of ecology (Figure 6.6a, b). Those who had teachers who were very committed to fieldwork were more likely to experience more open-ended tasks – such as the problem-solving activities advocated by many educationalists (Figure 6.6c, d). Identification has always been a particular problem, and this is underlined by the results of this research (Figure 6.6e). Not surprisingly, pupils' attitudes to fieldwork were clearly influenced by their teachers (Figure 6.7) and pupils who had been involved in problem-solving activities benefited more from the fieldwork experience (Gayford, 1985).

**The implications for higher education and research**
On the whole, those involved in higher education in Britain show little professional interest in what goes on in schools, certainly for ages 16 years downwards. This is not a criticism of the individuals working in that area. Very few people, other than a few admissions tutors, have a brief or a

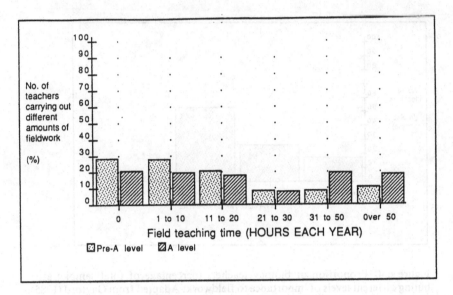

Figure 6.5. The amount of fieldwork carried out with different age groups in secondary schools (pre-A level refers to 11–16 year groups; A level to 16–18 years). Adapted from Fido and Gayford (1982).

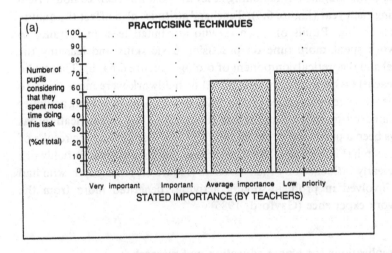

Figure 6.6. The influence of teachers' attitudes to fieldwork on the time spent on different fieldwork tasks. Based on self-assessment by pupils within different 'teachers' attitude' groups (see Figure 6.4). (a) Practising techniques, (b) Looking at ecological principles, (c) Problem-solving, (d) Measuring and counting, (e) Identification. Adapted from Gayford (1985).

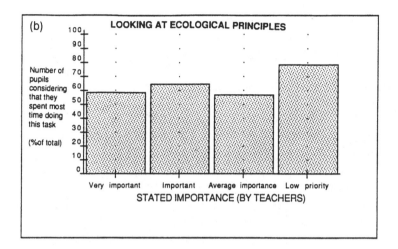

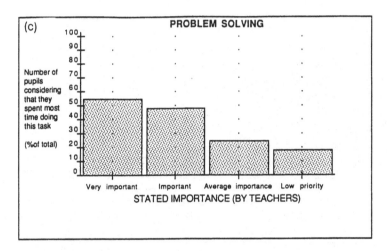

Figure 6.6. (*cont.*)

reason to keep up-to-date with educational developments in secondary education. An example of this schism was seen recently. As part of the introduction of a National Curriculum for 5–16 year olds in Britain, individuals and organisations were asked to respond to trial documents in different subject areas. University and polytechnic science departments showed very little interest in the most important educational initiative to affect science teaching in British primary and secondary education in nearly 50 years (e.g. Department of Education and Science, 1989b). Yet,

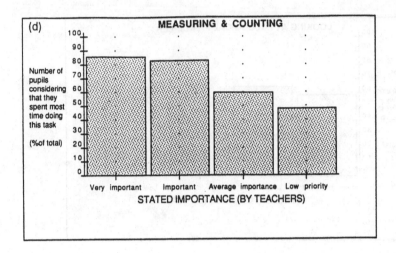

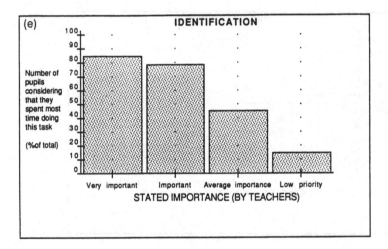

Figure 6.6. *(cont.)*

soon afterwards, at the 1990 meeting of the British Association for the
Advancement of Science, professional scientists were voicing concern
about the falling numbers of scientists emerging from secondary education. Their concern is well-founded. An example of how research can be
undermined by changes in education, both at school and college level,
already exists and has been well documented. Over the past 20 years (and
probably longer), teaching and research in taxonomy – the twin disciplines

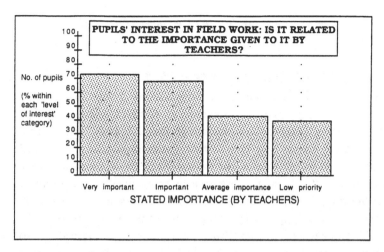

Figure 6.7. Numbers of pupils attributing at least a moderate degree of interest in fieldwork (as a proportion of the total number of students taught) within different 'teachers' attitude' groups (see Figure 6.4). Adapted from Gayford (1985).

of systematics and identification – has been severely curtailed by a combination of cuts in research funding and changes in the content of biology courses both at school and college level. The disappearance of trained taxonomists has now reached the stage where there are insufficient skilled taxonomists (including field biologists with good taxonomic skills) and resources (books, advisory services, etc.) to service the needs of ecological research (see e.g. NERC, 1976; House of Lords, 1991; Edwards and Walton, 1992). Whether this malaise started in secondary education, at undergraduate level or in research is uncertain. But it is undeniable that changes in course content and attitude in secondary schools have contributed to the undermining of one of the subjects that is fundamental to ecological research, at the very time when ecologists are calling for a much greater taxonomic input into contemporary work (e.g. Berry, 1988; Stork and Gaston, 1990). The 'cross-curricular' nature of ecology will always be a weakness (as well as a strength). Threats to 'service' disciplines will have to be fought as fiercely as direct threats to ecology itself, a fact acknowledged by Edwards and Walton (1992) when explaining why 'one science discipline (ecology) was so enthusiastically defending the importance of another (taxonomy)'. This is ironical when considering the reluctant and grudging acceptance of ecology by other scientists in the earlier part of this century (Tansley, 1951).

**What can be done to secure the future of ecology education in Britain?**

*Better integration of ecology into the curriculum*

Educationalists need to look closely at how ecology should be presented in the curriculum. Several models have been proposed whereby ecology would be taught as a theme running through the biology syllabus, rather than as a separate subject (e.g. King, 1980). Other authors have suggested that ecology could form the scientific basis of more environmental 'issues' based teaching, including a greater emphasis being given to human influences on the environment (Evans, 1988). This may result in the formerly exclusive role for teaching ecology being removed from the domain of the science or biology teacher; for example, giving a greater responsibility to geography teachers. Whatever happens, it is clear that the present situation causes confusion to all pupils (Harper, 1982). It is up to ecologists to show how the subject underpins much of what else is taught, both in biology and in environmental education generally.

*Higher profile for fieldwork*

Fieldwork is essential for teaching balanced ecology. Without it, pupils will not appreciate the exceptional problems, challenges and excitement presented by ecological work in the field and, furthermore, practical techniques will be taught inadequately. A move back into the classroom merely postpones responsibility until later in the education system – thus making fieldwork a privilege for those who are fortunate enough to go into higher education. This flies in the face of what most educationalists are striving for – a varied 'field experience' entitlement for all school children. The Department of Education and Science stated recently that 'Fieldwork in the country and city is an essential part of environmental education' (DES, 1989a). Furthermore, it should be remembered that today's pupils are tomorrow's teachers; a reduction in fieldwork experience will undoubtedly cause problems in the future teaching of ecology at secondary and graduate level. Compulsory inclusion of a fieldwork component in syllabuses would probably be the only way of ensuring that all pupils receive a full fieldwork entitlement. However, most ecological principles can be taught through laboratory experiments and classroom activities; this may entice those teachers who avoid ecology because of its association with fieldwork and the problems thereof to at least include classroom-based ecology in their schemes of work.

'Entombment' in the classroom or laboratory has the additional advantage of making assessment and the setting of exam questions easier – a factor not to be dismissed easily. The move indoors holds attractions for educational administrators and many teachers: for example, lower costs and easier timetabling are very seductive appeals. However, the total replacement of fieldwork by laboratory-based studies is an anathema to most ecologists. It is a move towards 'pot-filling' education. Field studies have provided the touch paper and the torch for the 'fire-lighting' education which has inspired many ecologists and environmentalists. But, it isn't sufficient for ecologists to invoke abstruse concepts such as 'well being', 'respect for the environment', 'better understanding', 'awe and wonderment' when advocating fieldwork. Hard facts and figures are needed as well; these exist, but they need to be communicated to those who make the decisions. For example, Herbert et al. (1972) suggested that field studies had a five-fold educational role. These are also true for the present day.

## (i) *Experience of the world*

All teachers have their own tales of how students' horizons have been widened. Barrett (1987), when describing the early days (1947 onwards) of a recently opened Field Studies Council centre, provides some salutory reminders of what can happen if biology is taught exclusively in rigid topics, and confined to the laboratory or classroom. He remembered 'The dear lady, seeing a large dogfish dead on the beach, insisted that it was not a dogfish. "If I may say so, Mr Barrett, I've been dissecting dogfish for more years than you have been alive" (and she had!) "and I can tell you that dogfish do not have a tail." Nobody had told her that tails were cut off to save postage. Or when looking at real specimens, a pupil exclaiming "Sir, sir . . . the book says . . . this animal is wrong."'

Many teachers who have taken classes into the field will have similar stories to tell.

## (ii) *Logical thought*

Fieldwork, by providing a wealth of first-hand data, can encourage both formal and intuitive problem-solving approaches. These are being actively encouraged by many educationalists (e.g. DES, 1989a, b) and provide a training that goes far beyond ecology. Unfortunately, as we have seen above, there is some evidence that problem-solving and other

fieldwork activities are not being developed as well as they could be
(Gayford, 1985).

### (iii) *Kindling enthusiasm (and respect for the environment)*

Some commentators have distinguished the 'pot-filling' and 'fire-lighting'
approaches to education. Fieldwork does have the capacity to inspire –
to instill a sense of awe and wonderment. Although it is not tangible
(and, therefore, easily discarded in the abstract profit and loss ledgers
of formal education) it is this aspect to which ecologists turn repeatedly
when asked to justify fieldwork. The 'pot-fillers' can also be comforted,
however, by evidence that shows that children can learn better and are
less distracted in novel surroundings – for example when comparing their
school to a nature centre (Falk, 1983). Other research has shown that early
'environmental' experiences by children can have a lasting effect on
attitudes (see Harvey, this volume). More anecdotal, is the fact that tasks
carried out in novel surroundings (such as field centres) persist in our
minds long after we have forgotten all details of the classroom-centred
courses.

### (iv) *Citizenship and environmental awareness*

One of the major developments in science and geography education at all
levels in Britain has been the introduction of environmental 'issues' into
syllabuses: topics such as conservation, pollution, energy conservation,
etc., now feature prominently. This is being supported, in principle at
least, by the government (Department of Environment, 1990). Experienc-
ing different environments must promote a greater awareness and
understanding of many of these issues and may provide a link between the
classroom and the wider world. In the absence of a broader environmental
experience the conceptual leap from activities in the classroom, laboratory and library to regional and global perspective is very demanding.
Under such circumstances, the path of logical development may have to
be traced by the teacher. With the additional experience provided by
fieldwork, particularly in novel environments, the extension from local
to regional (and global) perspectives is much more likely to be achieved
by the children themselves. The lessons will persist for longer. This is
recognised in several documents issued recently in Britain (e.g. DES,
1989a).

(v) *Technical training*

There are aspects of biology and ecology that can only be taught effectively in the field. Practical experience with techniques and equipment can only be achieved in 'real-life situations'. One of the main reasons why present-day teachers lack the confidence to teach field biology is the failure to provide adequate training in the field during their years as students. A further dilution, or elimination, of the fieldwork component in biology syllabuses will serve to aggravate the situation. The consequence for higher education will be that it will have to take on the role of training biologists with no previous field experience. Otherwise, the likely result is a dearth in qualified and experienced graduates at a time when the demand for such expertise, from industry as well as traditional sources (conservation bodies, etc.), is growing. These arguments may appear flimsy but a stark reminder of what can happen if sufficient notice is not taken of changes in secondary education has already been demonstrated by the decline in taxonomy as discussed above.

Reassuringly, there seems to be little support for the abolition of a fieldwork component in the curriculum. Environmental Education, in which ecology must play a prominent role, is being actively promoted as part of numerous educational initiatives (e.g. European Commission, Council of Education Ministers, 1988; DES 1989a; DoE, 1990). Most of these stress the importance of field studies. But these initiatives are worthless if sufficient resources and facilities are not made available to carry out the task. It is for these reasons that there is a real concern that fieldwork may actually be declining in schools.

### Teacher-training

It is obvious that many biology teachers are still intimidated by the prospect of 'fieldwork'. Recently introduced safety guidelines have served to reinforce this insecurity (DES, 1989c). This will reduce the amount and quality of fieldwork being taught in our schools. This can only be overcome by providing adequate experience and training at all levels of education. It is not sufficient to target student-teachers. There is evidence to suggest that older teachers have the lowest regard for fieldwork (Fido and Gayford, 1982). In-service training courses in field ecology must become a priority, but compulsory inclusion of a fieldwork component in syllabuses will be the only measure that will ensure a radical improvement.

*Facilities and resources*

Sufficient funding must be provided to maintain the present level of facilities and resources. Recent studies (CEE, 1992) suggest that this is not happening. Commerce, industry and higher education could play a more prominent role; they are, after all, the ultimate consumers. The Field Studies Council's AIDGAP (Aids to Identification in Difficult Groups of Animals And Plants) project is a good example of what can be achieved with minimal funding (see Tilling, 1987, for further information). This project has attempted to diminish the difficulties of identification, an area that is seen consistently as one of the major problems facing inexperienced teachers. In 15 years, AIDGAP has distributed over 40 000 guides to schools and individuals throughout Britain. AIDGAP now finds itself under threat because its funding is coming to an end. Although organisations such as the British Ecological Society have offered partial funding, approaches to industry and government departments have met with a disappointing response.

*Research into the present state of ecology education and fieldwork*

Most of the formal research into ecology education was carried out over ten years ago. There is an urgent need for more up-to-date research to find out what is happening in schools today – particularly investigating the impact of new safety guidelines and changes in education such as the introduction of the National Curriculum and other legislation introduced as a result of the Education Reform Act 1988.

*A clearer career structure*

In Britain, there is no professional standard for ecologists and virtually anyone can call themselves an ecologist, and many people do. This has devalued 'ecology' as a recognised profession. There is no professional qualification for people working in such fields as environmental impact assessments, environmental management and conservation. There is no defined career structure for ecologists and this may have a significant, but under-rated, impact on the numbers and quality of students choosing biology and environmental science options, particularly at undergraduate level. However, in 1991 the Institute for Ecology and Environmental Management was launched. One of its main purposes is to establish recognised professional standards in ecology and environmental management and, eventually, to bestow chartered status on qualified individuals.

The inception of such a body may help to secure the future of ecology as a profession.

### Public pressure

Scientists tend to underestimate the power of public pressure and the value of publicity. For example, in 1980 the British Ecological Society (BES) published a notice asking members to help in answering an 'overflowing postbag of enquiries from the public' (particularly on pollution and conservation), many from 'schoolchildren engaged on project work'. Just 16 members (out of the then membership of 3512) volunteered (Sheail, 1987). Such insularity compared unfavourably with other countries and would have been damaging if allowed to persist. However, like its counterparts in other countries (see Berkowitz, this volume) the BES now has an education section – the Teaching Ecology Group (Sheail, 1987). It is important to recognise that the major catalysts of the recent radical changes in the British education system, under the auspices of the Education Reform Act 1988, were concern by the general public, often expressed through the media, and commercial interests, who were worried about the standard of their intake from education. Publicity, possibly focussing on the need for rigorous and extensive ecological data to support work on high profile environmental 'issues', could become a powerful weapon in the ecologist's armoury.

### Conclusions

Present-day education in Britain is undergoing a number of radical changes. These changes should be exploited to promote ecology education. In particular, the increasing importance being given to environmental education and out-of-classroom experience, along with the emphasis on problem solving, data handling and observational skills, provides an opportunity for ecology to acquire a higher profile in the curriculum; after all, the discipline provides the fundamental scientific base that underpins the much wider theme of environmental education. However, there are many demands being placed on the education system and some are in conflict with, and could undermine, the provision of 'good' ecology teaching. However, the rewards will only come to those who ask. Ecologists must press for a wider acceptance of ecology's role as a scientific subject and as a vehicle for teaching a whole range of skills being demanded in other areas of the curriculum. This may mean that some ecology will be taught in subject areas other than biology or geography, but it should not be severed from its scientific base. The impetus for such changes in schools should come from all areas of education, including higher education, and from the growing number of 'external' interests in ecology. If the present-day opportunities are allowed to disappear then much of what is considered 'good' ecology teaching may become diluted or eliminated from British education.

## References

Arthern, M. (1985). Fieldwork in Secondary Schools: A survey and some consequences. *National Association of Field Studies Officers Journal*, **10**, 32-39.

Barrett, J.H. (1987). The Field Studies Council: how it all began. *Biological Journal of the Linnean Society*, **32**, 31-41.

Berry, R.J. (1987). Scientific natural history; a key base to ecology. *Biological Journal of the Linnean Society*, **32**, 17-29.

Berry, R.J. (1988). Systematics in the education of biologists. In *Prospects in Systematics*, ed. D.L. Hawksworth, pp. 414-429. Oxford: Clarendon Press.

Berry, R.J. (1989). Ecology: where genes and geography meet. Presidential Address to the British Ecological Society, December 1988. *Journal of Animal Ecology*, **58**(3), 733-760.

Booth, P.R. (1979). The teaching of ecology in schools. *Journal of Biological Education*, **13**(14), 261-266.

Cherrett, J.M. (1989). Key concepts, the results of a survey of our members' opinions. In *Ecological Concepts*, ed. J.M. Cherrett, pp. 1-16. Oxford: Blackwell Scientific Publications.

Council for Environmental Education (CEE) (1992). *Local Authority Provision for Environmental Education*. Reading: Council for Environmental Education Report.

Crothers, J.H. (1987). Formative fieldwork: the age of the sixth form course. *Biological Journal of the Linnean Society*, **32**, 49-57.

Department of Education and Science (1989a). *Environmental Education from 5 to 16. Curriculum Matters 13*. London: HMSO.

Department of Education and Science (1989b). *Science in the National Curriculum*. London: HMSO.

Department of Education and Science (1989c). *Safety in Outdoor Education*. London: HMSO.

Department of Environment (1990). *Our Common Inheritance*. London: HMSO.

Edwards, P.J. and Walton, D.W.H. (1992). The state of taxonomy: an ecologist's view. *Bulletin of the British Ecological Society*, **13**(1), 17-26.

Environment Council (1990). *Who's Who in the Environment*. London: The Environment Council.

Evans, S.M. (1988). Man and the environment: the need for a more realistic approach to teaching ecology. *Journal of Biological Education*, **22**(2), 136-138.

European Commission (1988). *Statement by Education Ministers.* EC Resolution 88/C177/03.

Falk, J.H. (1983). Field trips: A look at environmental effects on learning. *Journal of Biological Education,* **17**(2), 137–142.

Fido, H.S.A. and Gayford, C.G. (1982). Field work and the biology teacher: a survey in secondary schools in England and Wales. *Journal of Biological Education,* **16**(1), 27–34.

Gayford, C.G. (1985). Biological fieldwork – a study of the attitudes of sixth-form pupils in a sample of schools in England and Wales. *Journal of Biological Education,* **19**(3), 207–212.

Harper, G.H. (1982). Why not abolish ecology? *Journal of Biological Education,* **16**(2), 123–127.

Herbert, A.T., Oswald, P.H. and Sinker, C.S. (1972). Centres for field studies in England and Wales: the results of a questionnaire survey in 1969. *Field Studies,* **4**, 655–679.

House of Lords (1991). *Systematic Biology Research.* Select Committee on Science and Technology, House of Lords. HL Paper 41. London: HMSO.

King, T.J. (1980). *Ecology.* Walton-on-Thames: Nelson.

King, T.J. (1989). Alternative ecology. *TEG newsletter,* **5**, 2–3.

Natural Environment Research Council (1976). *The Role of Taxonomy in Ecological Research.* NERC Publications Series B. Swindon: NERC.

Pantin, C.F.A. (1968). *Relations Between the Sciences.* Cambridge: Cambridge University Press.

Sheail, J. (1987). *Seventy-Five Years in Ecology: The British Ecological Society.* Oxford: Blackwell Scientific Publications.

Southwood, T.R.E. (1977). Habitat, the template for ecological strategies? *Journal of Animal Ecology,* **46**, 337–365.

Southwood, T.R.E. (1988). Tactics, strategies and templets. *Oikos,* **52**, 3–18.

Standing Conference on University Entrance (1977). *Knowledge of Biology Required by University Departments Specializing in Biological Disciplines.* London: SCUE.

Stork, N. and Gaston, K. (1990). Counting species one by one. *New Scientist,* **1729**, 43–47.

Tansley, Sir Arthur (1951). What is ecology? Council for the Promotion of Field Studies, reprinted in *Biological Journal of the Linnean Society,* **32**, 5–16, (1987).

Tilling, S.M. (1987). Education and taxonomy: the role of the Field Studies Council and AIDGAP. *Biological Journal of the Linnean Society,* **32**, 87–96.

# 7

# Field studies as a technique for environmental education in developed and developing nations

WALTER D.S. LEAL FILHO

*Principal Research Fellow, Department of Environmental Sciences, University of Bradford, UK*

## Introduction

One of the primary objectives of ecology teaching is the development of a sense of environmental awareness in young people: a goal which implies fostering their environmental education. It is widely accepted that in terms of ecology teaching, contact with soil, water and organisms helps pupils to understand ecological processes and to become aware of the dynamics of the environment (see Harvey, Chapter 8). This contact with the elements and with living things has its didactic potential maximised by field studies.

The evolution of field studies is closely related to outdoor pursuits as accessory techniques of environmental education. However, even though this is a global trend, seen in both industrialised and semi-industrialised nations, its development is uneven.

Assessment of conceptual and factual aspects of field studies and their level of implementation in both developing and developed countries requires considerable research. This paper aims to describe the evolution of field studies in a developed country, in this case, Great Britain, compared with a developing nation, namely Brazil. As a further illustration of the conditions in both types of countries, results of a survey of field studies undertaken in Britain and Brazil are presented with suggestions for change in developing countries.

## Brief history of field studies: the British example

Britain has a long tradition of field studies and its complex growth as an educational strategy has been emulated to a greater or lesser extent worldwide. The British example will be described to show how field studies have evolved as an educational tool.

Records of field studies in Britain as part of formal teaching go back to 1888 when, according to Nicholson (1968), Sir Jonathan Hutchinson founded the Haslemere Educational Museum. Later records indicate the role played by the National Trust for Places of Historic Interest and Natural Beauty, founded in 1895, which had its first site at Dinas Olen, a cliff near Barmouth in North Wales (Parker and Meldrum, 1973). The site was frequently visited by teachers and school children to undertake field studies. Subsequently there were a number of initiatives to stimulate the use of the natural environment in education, including, among others, activities in field studies conducted by the Royal Society for the Protection of Birds (RSPB) with school children in 1904 and by the Association of School Natural History Societies in the same decade.

With the establishment of the Forestry Commission in 1919 the didactic potential of forests in practical work was realised. Initiatives stimulating the use of field studies were strengthened by the formation in 1926 of Britain's first County Naturalists' Trust in Norfolk. In addition, the Council for the Preservation of Rural England was created a short time later. Both institutions included in their plans of action the use of environmental resources in the countryside in conservation education. The first 'methodological' initiative for conservation education in natural areas took place in Lindsey – near Ipswich – when, as a result of the need for enforcement of the Sandhill Act of 1930, conservation programmes, including initiatives to stimulate the awareness of the general public and school children, were established in order to avoid the misuse of coastal sand dunes and their surrounding area.

In 1930, with the growing use of the outdoors in teaching as well as for leisure, the Youth Hostel Association was created to promote the use of natural areas in a range of leisure and fieldwork activities. In the following decade, the Field Studies Council (FSC) was established. The FSC opened its first field study centre at Flatford Mill in 1946. A number of other centres followed and now the FSC owns ten centres in England and Wales.

The Nature Conservancy Council established the 'Study Group on Education and Field Biology' (SGEFB) in 1960. Amongst other duties it examined the role of field studies and their relation to school education in general and science teaching in particular (SGEFB, 1963). This major assessment of field studies in formal teaching in Britain provided a significant contribution to the development of field studies as a methodology for environmental education.

During the 1960s the growing use of in-service training programmes for teachers in the use of the environment in teaching gave a boost to the use of the outdoors in practical work. The Nuffield Foundation co-ordinated the Junior Science Project and Nuffield Science Teaching Project (including the Nuffield A-Level Biology Trials). These initiatives led to an increase in students' achievements, including the development of practical abilities, as a result of work conducted in the field (Dowdeswell 1967; Selmes, 1973; Kelly and Nicodemus, 1973). In addition to its direct contribution to the development of practical studies within the school syllabus the Nuffield Biology Project is acknowledged to be the first introduction of the principles of ecology in lower and upper secondary schools.

In describing the evolution of field studies, mention should be made of the contribution by the Outward Bound Schools Project, which aimed at exposing young people to testing experiences in order to discover their personal reactions and limits (Fletcher, 1970). This programme was similar to other schemes, such as the Brathay Hall Trust and the Duke of Edinburgh Award Scheme, which over the years have stimulated the development of outdoor pursuits in general and field studies in particular.

In March 1965 a conference on education in the countryside held at Keele University concluded that the implementation of specific programmes of environmental education in both formal and non-formal education was desirable, including the development of field studies. The need for a co-ordinating body responsible for the development of environmental education programmes was also expressed. The following conclusions were reached by the Conference:

1. Positive educational methods were needed to encourage awareness and appreciation of the natural environment as well as responsibility for its trusteeship by every citizen, especially by school children, who will be occupying key positions in society in the future.
2. The education system has a decisive contribution to make in creating this awareness and sense of individual responsibility.
3. The countryside is a rich source of inspiration and teaching material, which can contribute substantially to education at all levels, field studies provide a valuable means of using and developing this educational resource.
4. If properly integrated into the curriculum, field studies can form a valuable part of the whole school curriculum. They provide an opportunity to relate many subjects directly to the natural environment and foster a deeper understanding of the forces affecting it.

Thus by the mid 1960s field studies had achieved a high level of acceptance as an environmental awareness-raising strategy. In November of the same year as the conference at Keele a conference organised by the FSC reached conclusions similar to the ones drawn from Keele, particularly regarding field studies. This event, the 'Conference on field studies at residential centres' emphasised the role played by field studies in the development of a sense of environmental awareness among students, parallel to the approach of curricular topics.

As a result of the recommendations from the Keele Conference, the FSC Conference and other initiatives, the Council for Environmental Education was established in 1968. The Council was expected to act in England and Wales. In 1969 the Committee on Education in the Countryside (now called the Scottish Council for Environmental Education), was founded to act in Scotland. Both organisations have reinforced the movement for environmental education as a whole and field studies in particular in Britain.

Over the last 20 years field studies have evolved from an educational strategy used by a few teachers to an established methodology. In the late 1960s, with the widespread use of field studies centres in Britain and the rest of Europe, the increasing popularity of fieldwork led to conferences, courses and seminars aimed at preparing teachers to approach environmental issues in the outdoor environment. Kerr (1963) showed the importance of practical work in the teaching of science, emphasising the need for pupils to have first-hand contact with nature.

There has been a steady growth in the training and preparation of personnel to develop field studies and an increase in the use of outdoor education centres and field studies centres.

## Concepts of field studies

In the context of the present study, field studies are characterised by the integration of practical activities in the outdoor environment as part of a range of subjects. Field studies differ from 'outdoor education' as the latter is not always related to the school curriculum.

The definition of outdoor education provided by the National Association of Outdoor Education (NAOE) states that 'outdoor education is a means of approaching educational objectives through direct experience in the outdoors, using as learning materials the resources of the countryside and the coastline' (NAOE, 1972), i.e., studies of the environment (in the context of teaching subjects such as biology, geography or environmental sciences, etc.) and outdoor pursuits not related to subjects' contents (e.g. canoeing, mountaineering and rock climbing).

Morine (1983) defined field studies as 'a process of learning and teaching about the environment that may be conducted within the context of formal teaching through direct contact with the ecosystems, fauna and flora as well as with natural resources'. Thus field studies are closely related to subjects' contents, having a more constant presence in the process of formal teaching. They provide practical experience, which is essential to relating to the real world the contents of subjects that otherwise would be abstract and easily forgotten by the students.

One of the most valuable characteristics of field studies in formal teaching is that they can be easily included in the context of subjects such as biology, history, geography, environmental sciences, environmental studies, ecology, science, geology and social studies. This characterises a major feature of field studies, namely their *interdisciplinarity*. Interdisciplinarity is defined by the Centre for Environmental Research and Innovation (CERI, 1972) as describing the interaction between two or more different disciplines.

Field studies represent educational strategies based on the use of nature and natural resources as teaching tools, which are conducted through a combination of the knowledge offered by different subjects, and applied during the development of practical activities. They stimulate pupils to have an appreciation of the elements: air, soil, water, and organisms. Field studies also represent an effective means through which a sense of commitment to environmental conservation may be passed on to pupils.

It can be seen from the definitions of outdoor pursuits and field studies that clear boundaries exist between the two. Although both refer to living and learning outside classrooms, field studies are closer to the teaching process. As a result there has been a search for the development of techniques to maximise the didactic potential of field studies, based on their importance in the process of formal education of school children. Within the conceptual analysis of field studies it is also relevant to mention the SGEFB's (1963) assessment of its relevance to a child's education:

It may be that a child's innate interest in animals and plants, earth, air and water can be caught up in new ideas to lead to an understanding of scientific principles and methods, and of the uses of experiment and the making of exact records and identification. In so far as this is true, field studies could become, especially for young children, the gateway for science as a whole.

According to the National Association of Field Studies Officers (1989), the value of fieldwork as a methodology of environmental education resides on the fact that it:

- provides an essential part of the cross-curricular environmental approach to both core and foundation National Curriculum subjects;
- gives relevance to topics which could otherwise remain as matters of second-hand learning;
- provides pupils with opportunities to talk about their own work, to listen to others and share their knowledge;
- is concerned with real people, real situations and real issues.

In addition, field studies encourage both self-discipline and courtesy (Morgan, 1982) as children are required to respect regulations.

The conceptualisation of field studies also involves a description of its objectives, which are an improvement in the relationship of humans to nature through practical experience. There has been some debate on the aims of field studies in the context of environmentally related subjects by a number of workers (including Masterton, 1973; Sinker, 1979; Webster, 1981). From an investigation of the literature and an analysis of the definition and benefits of field studies the following objectives of field studies recur:

- to develop environmental awareness and an understanding of the complexity of the environment;
- to maximise the potential of the environment in education;
- to integrate knowledge from different subjects based on the principle of interdisciplinarity and the adoption of an holistic approach;
- to stimulate logical thought via solving fieldwork related problems using a scientific approach;
- to stimulate cognitive and affective values enhancing pupils' moral and social virtues.

The combination of these factors is acknowledged as leading to a better understanding of environmental dynamics and in the preparation of pupils for adult life. One of the most important conceptual characteristics of field studies is that such a methodology involves the application of skills and adoption of behaviours not easily registered in the classroom. Positive attitudes and values are developed through this strategy, as are a sense of environmental awareness, which takes place parallel to the evolution of students' personal and social education.

**Field studies as a tool for environmental education**

As mentioned earlier, as a result of the growth of environmental education on a worldwide scale there has been an intensive search for techniques and methodologies through which its didactic potential may be maximised. In biological and environmental sciences, ecology and other environmentally-related subjects, the integration of field studies and environmental education is essential.

In the case of biology for example, it was demonstrated by Perring (1969) that an increase in the number of field biologists is needed to lead students on field visits and to demonstrate the dynamics of nature. This does not necessarily mean that the person should be a biologist. The learning of the principles of biology and practices of field studies by mathematics, history and geography teachers needs to be developed to a greater extent. A considerable amount of work exists on a wide range of environmental education topics related to the use of fieldwork within formal teaching.

Some studies conducted in natural areas or involving the use of the natural environment may be adapted to develop approaches to environmental education. For example, Shepley and Coultas' (1971) experimental work on animal behaviour involving herring gulls (*Larus argentatus*) can be adapted for other animal species threatened by extinction; Dalby's (1970) programme of plant taxonomy provides a good opportunity to introduce to students the diversity of form found in nature through classification; Prosser's (1982) use of the demonstration laboratory combined with the collection of samples and data in the field can be used to aid biology teaching; and Caulton's (1970) ecological approach to biology, emphasising ecological issues can also be used.

However, there is still a lack of research as to how field studies can be used to engender environmental awareness among pupils, how this can contribute to pupils' environmental education, and how this can be integrated in the context of the curriculum. Some examples detailing the environmental education potential of field studies are evident from the literature: Clarke (1967) for example, investigated the value of field studies for 11–15-year-old pupils and found that:

1. Tests given to the students showed that in terms of effectiveness field studies were superior to the corresponding indoor classroom work.
2. Field studies enhanced the ability of students to write and to express their own ideas logically.

3. The majority of students who participated in the experiment enjoyed it and found the work a pleasant experience.
4. There was a stimulation of group sense and an increase in awareness of the importance of joint work.
5. Younger pupils (11–12 and 12–13 years old) seemed to benefit more from field studies than the older ones.
6. There was an increase in pupils' appreciation of nature and natural resources.

This last finding of Clarke's research is possibly the most significant in terms of showing the didactic potential of field studies in the area of environmental conservation. The environment is not always aesthetically pleasant to study due to physical deterioration or pollution. For example, high levels of pollution in a city or in a river provide an opportunity for pupils to develop an awareness of human influences on the environment.

Field studies contribute to environmental education by:

1. providing first-hand experience about the natural environment. For this to be effective activities need to be carefully planned and organised with preparation and follow-up procedures well established before work commences.
2. facilitating the discovery of inter-relationships between the physical, biological and social environments. Both pupils and teachers need to integrate the skills that are derived from different sources and subject areas. In carrying out field work, the assimilation of knowledge from subjects such as biology, geography, geology, chemistry, physics and social studies is often required.
3. motivating an appreciation of, and a sense of commitment to, the conservation of the environment, by facilitating an understanding of the structure of ecosystems and the interaction of social, economic and political aspects.

In addition, field studies stimulate a positive sense of moral and social behaviour that should be promoted in community life, making pupils more active and conscious of their role as citizens. This awareness is fundamental in developing environmental conservation activities, leading to the lobbying of governmental and non-governmental institutions for change whenever their actions militate against conservation or the balance of ecosystems. These advantages add value to indoor instruction, encouraging pupils to view the environment with a critical eye, combined with their natural curiosity.

**Field studies and environmental education in developed nations: a case study of England and Wales**

Mainland Great Britain, comprising England, Wales and Scotland, covers an area of 244 103 square kilometres and supports a population of around 57 million people. The distribution of inhabitants is not even, the southern part of England and Wales being more densely populated than other regions.

Approximately 1% of the British population are illiterate (Mazzolenis, 1989), primarily found in the adult population as attendance at schools is compulsory for children from 5 to 16 years old. There are approximately 9 million children of school age in Britain and the level of school attendance is one of the highest in Europe. There are approximately 35 913 primary and secondary schools with an entrance of 9 394 841 (1987), excluding Northern Ireland. Great Britain allocates approximately 11.0% of its GNP to education.

These figures, when considered alongside the development of environmental education and field studies in Britain (presented earlier), indicate that a high level of resources are invested in education in Great Britain. A significant number of environmental education programmes are being conducted and attention is also being drawn to the methodological development of environmental education. Environmental education programmes are generally undertaken by individual institutions or by joint initiatives involving two or more organisations.

Work of a more academic nature in environmental education has been carried out by, for example, the Council for Environmental Education (CEE, 1984), Parry (1987), Gayford (1986a), and others. These studies provide information enabling a better understanding of environmental education practices, assist future research initiatives, and guide government led environmental education initiatives. In addition, numerous sources of information on environmental issues has resulted in extensive published material on environmental education.

Another indicator of the extent to which environmental education is widespread in Britain is the proliferation of non-governmental institutions (associations, societies and trusts), concerned with environmental conservation. These help to promote the principles of rational use of the environment throughout the country. Several environmental education programmes have sought to engender a sense of environmental awareness among pupils. Research such as that of Fido and Gayford (1982), Gayford (1986b) and NAFSO (1989) has evaluated this growing trend. The amount and diversity of environmental education in formal and non-formal teaching in England and Wales (see DES, 1982 and CEE, 1984) is among the highest in Europe. This is supported by numerous technological and methodological developments in the UK, particularly those of the 1970s and 1980s.

## Field studies and environmental education in developing nations: a case study of Brazil

By contrast, environmental education in developing countries is not as far advanced as that in the developed world. The situation in Brazil is summarised to give an indication of the status and progress of environmental education in general and field studies in particular.

Brazil is the largest South American country, and the fifth largest in the world, with an area of 8 511 965 square kilometres. Brazil is the eighth largest economy of the capitalist nations and has the eighth largest population in the world, with 155 million (1990). Brazil's demographic concentration is far lower than Britain but has a much higher rate of population growth, which in the period 1980–1985 was approximately 2.3%. Nearly half the population is under 25 years old.

Brazil occupies 74.3% of South America, and is the continent's most populated country. Its inhabitants are irregularly distributed: the highest concentration of population is along the coast line and in the southern states of Minas Gerais, Rio de Janeiro and Sao Paulo. The large urban population is exacerbated by constant internal migration from the north to the south where most of the industry is located. This causes problems of unemployment, violence and difficulties in providing housing in the major cities and the creation of 'favelas', or shanty towns, now a permanent problem.

The Brazilian teaching system has a structure similar to that of Britain (Leal Filho, 1991), being divided into three sectors: primary, secondary and tertiary. There are 211 785 primary and secondary schools in the country, with an entrance of 30 160 240 students in 1988. A recent official survey conducted by Mazzolenis (1989), indicated that approximately 25% of the Brazilian population is illiterate. In 1988 Brazil spent 6.7% of its GNP on education. Despite the constitutional requirement to educate children from 5 to 16 years old, approximately two million children of school age do not attend school due to lack of schools or the inability of parents to purchase teaching materials. In addition, the dropout level is very high, especially in rural areas, where children are usually taken from school to help their parents in agriculture.

Most of the successful initiatives in ecological education are currently developed by private institutions and professionals in Brazil (there *is not* as yet a formal programme of environmental topics within the curriculum). These initiatives are aimed at raising environmental issues within separate subjects. To date, there is no register of initiatives in practical activities (such as field studies) in the non-official programmes of environmental education. Initiatives stimulating the use of fieldwork in environmental education projects are inhibited by a lack of research into methodological and instructional aspects of conducting field studies as well as a deficiency in the number of relevant publications. These deficiencies were listed by Leal Filho (1991) in an analysis of the evolution of and current trends in environmental education in Brazil.

The author noted that a lack of publications about the environment, among other factors, is a serious impediment to the development of environmental education in Brazil. Few publications are available in science education dealing specifically with environmental topics. In addition, there is a scarcity of publications on methodological and instructional aspects of environmental education as well as on the organisation of field studies. This illustrates the difference between the current status of environmental education practices in the developed world and that in developing countries. The under-development of field studies in developing countries results primarily from the lack of the appropriate educational elements in the curriculum (both in terms of content and in the context of teaching programmes and learning styles). These cannot easily be implemented in the teaching systems of developing nations. The development of new initiatives to enable the integration of field studies in teaching programmes is needed in these countries.

**Field studies in Brazil and in Britain: a comparative study**

The value of field studies as an approach to environmental education in formal teaching is unquestionable in both developing and developed countries. However, as the above comparison between Britain and Brazil illustrates, the disparity between these countries in terms of the development and provision of field studies is great.

In this comparative study of the development of environmental education in an industrialised and a developing country, Britain was selected for study as considerable emphasis is placed on environmental education in the formal and non-formal sectors of education and a range of academic discussion on the development of field studies has occurred. It was considered that British field study practices could contribute a positive input to the establishment of such strategies in a developing country such as Brazil, which suffers from a number of environmental problems (Leal Filho, 1987) and desperately searches for effective environmental education techniques.

Based on such needs, a questionnaire survey of field studies was carried out between 1987 and 1989 in a sample of secondary schools in all 26 Brazilian states and in 45 British counties. This included data collected from head teachers, teachers and students, complemented by a number of visits and interviews. The study also aimed to assess the status of field studies as part of formal teaching in the countries investigated; to quantify the number of teachers with training in field studies; to gather teachers' and students' opinions on the sort of work undertaken; and to suggest, based on the study's findings, measures that could be adopted by the Brazilian authorities in order to maximise the effectiveness of field studies in that country.

## Results of the survey

The study showed that British schools undertake three times more field studies than Brazilian schools (Figure 7.1). In both countries internal differences were registered: in Britain most of the schools that indicated the existence of an annual field studies programme were in the south of

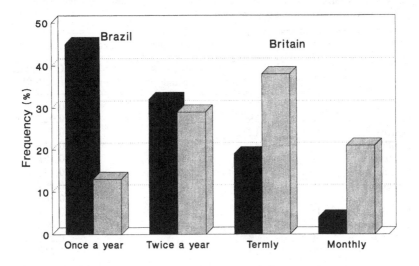

Figure 7.1. Annual frequency of field studies in Brazil and Britain.

England, while in Brazil such schools were in the south-eastern region of
the country. Regarding the areas used for field studies, in Britain the
number of sampled schools undertaking field studies in areas other than
school grounds is relatively high (over 70%), while in Brazil they account
for only 43%. Areas at considerable distance from schools or away
from cities are seldom used for field studies in Brazil, and no fieldwork
activities are undertaken at field studies centres as such institutions do not
exist. In Britain over 60% of schools investigated possess a vehicle
whereas only two of the schools surveyed in Brazil have such a resource.

Differences are also evident in teacher training. The survey found that
Britain has a far wider and more organised teachers' training programme
than Brazil: 69% of the sampled teachers in Britain attended a training
course enabling them to carry out field studies, a response given by only
26% of the Brazilian teachers surveyed. Such training among Brazilian
teachers, as shown in Table 7.1, is concentrated among tutors of science,
biology, environmental science and geography, and is virtually absent
among teachers of other subjects.

The survey also showed that the period during which field studies are
undertaken throughout the year differs in both countries (Figure 7.2): in
Brazil September and October are the peak months, in Britain it is fairly
regularly distributed from April until August.

Table 7.1. *Numbers of Brazilian and British teachers trained in environmental
issues, according to teachers' statements*

| Subject | Total of teachers | | Number of trained teachers | | Chi-square $p < 0.05$ |
|---|---|---|---|---|---|
| | Brazil | Britain | Brazil | Britain | |
| Science | 24 | 22 | 8 | 18 | 0.045 |
| Biology | 68 | 38 | 21 | 25 | 0.022 |
| History | 17 | 12 | 0 | 9 | 0.001 |
| Geography | 54 | 47 | 26 | 38 | 0.016 |
| Geology | 16 | 17 | 3 | 14 | 0.101 |
| E. Science | 6 | 26 | 2 | 20 | 0.138 |
| Languages | 10 | 11 | 0 | 4 | 0.001 |
| Arts | 11 | 12 | 3 | 11 | 0.106 |
| Maths | 11 | 10 | 0 | 2 | 0.001 |
| Chemistry | 14 | 7 | 0 | 5 | 0.001 |
| Physics | 9 | 8 | 2 | 2 | 0.250 |
| Humanities | 20 | 15 | 3 | 7 | 0.119 |

Brazil, $n = 260$; Britain, $n = 225$.

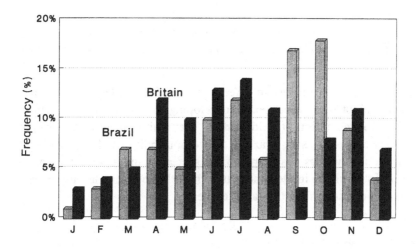

Figure 7.2. Frequency of field studies throughout the year according to pupils' statements.

The research also indicated that despite the socio-economic differences between these countries and the unequal development of field studies (only 32% of the sampled students in Brazil affirmed having done field-work in the period 1988/1989 compared to 83% of British respondents), British and Brazilian pupils have similar attitudes towards field studies: they were largely positive. Finally, in both Britain and Brazil, students' high receptivity to field studies shows that they can significantly benefit from the range of knowledge, understanding and skills engendered through fieldwork activities.

## Conclusions

The comparative analysis of field studies in Britain and Brazil revealed that field studies in the latter are unequally undertaken, unsystematically developed and uncoordinated. As a result of these deficiencies initiatives by individual schools are of little overall effect. As in other developing countries field studies need to be formally integrated in the teaching system in Brazil. To reach this goal, the Brazilian example shows that three main requirements need to be fulfilled:

1. The need for an in-depth study of the curriculum, including analyses of the contents of all subjects, in order to develop strategies to incor-porate field studies in the context of the different disciplines.

2. The provision of teachers' training programmes to develop environmental topics and promote field studies in formal teaching.
3. Official support, both logistical and financial, for the implementation of field studies in schools.

The British education system offers a range of examples of procedures which if adapted to the context of developing countries would be helpful in the implementation of field studies. The establishment and oversight of education policies in Brazil is the responsibility of the Ministry of Education. Key measures adapted from successful initiatives taking place in Britain should be implemented at each of the following levels: government, school administration and teachers.

### (a) Government level

1. Establishment of a working group of teachers, scientists and experts in education from different disciplines; the working group should establish a nation-wide field studies curriculum with proposed activities for different subjects and should oversee its implementation.
2. Inclusion by the working group of elements incorporating environmental concepts and promotion of environmental values, attitudes, knowledge and skills in the context of individual subjects.
3. Government commitment to promote pedagogic as well as scientific research into methodologies of field studies, including analysis of field studies practices and evaluation of the ways that different ecosystems may be used in practical activities.
4. Provision of financial resources from the Ministry of Education for the implementation of field studies, and for the production and diffusion of instructional and audio-visual materials for use in teaching environmental issues and as background materials for field studies.
5. Provision of funding for the inclusion of environmental inputs, including field studies, in teacher education courses and in-service training schemes.

### (b) School level

1. Each school should have its own annual budget, to enable field studies programmes to be implemented: to provide transport of students and field studies materials.
2. Head teachers should ensure their staff undertake field studies and take part in the preparation of field studies programmes. They should

also provide formal support, supervise the planning of fieldwork activities and ensure proper logistical back-up is available.

3. As the importance of field studies may not be apparent to all teachers, head teachers should ensure their staff participate in training schemes.

### (c) Teacher level

1. Teachers from different disciplines should plan the inclusion of field studies, particularly when environmental issues are raised in the context of their subjects.
2. Assessment schemes should take into account students' performance in field activities and not be just on the basis of classroom work.
3. A structured approach is required to ensure the proportion of theoretical work undertaken in the classroom is balanced with field based experience.

The recommendations above provide a starting point for the successful implementation of field studies in Brazil and in other developing countries. The British experience shows that field studies are a complementary approach to environmental education. The goal of preparing individuals for a harmonic relationship with the environment must be viewed as an investment for the future well-being of communities and of all nations.

### References

Caulton, E. (1970). An ecological approach to biology. *Journal of Biological Education*, **4**, 1–10.

Centre for Environmental Research and Information (CERI) (1972). *Interdisciplinarity: Problems of Teaching and Research in Universities.* Paris: OECD.

Clarke, J.H. (1967). An investigation into the value of field studies to pupils of 11–15 years old. *School Science Review*, **49**, 11–18.

Council for Environmental Education (CEE) (1984). *Improvements in Information Provision for Environmental Education in the United Kingdom.* A Report Prepared by the Council for Environmental Education for the Department of Environment. Reading: CEE.

Dalby, D.H. (1970). Plant taxonomy as a field study. *Journal of Biological Education*, **4**, 115–132.

Department of Education and Science (DES) (1982). *Environmental Education: Sources of Information.* London: HMSO.

Dowdeswell, W.H. (1967). The Nuffield biology project at 'O' Level. *Journal of Biological Education*, **1**, 29–37.

Fido, H. and Gayford, C. (1982). Fieldwork and the biology teacher: a survey in secondary schools in England and Wales. *Journal of Biological Education*, **16**, 27–34.

Fletcher, B.A. (1970). *Outward Bound – A Follow-up Study.* Department of Education, Bristol University.

Gayford, C.G. (1986a). Biological fieldwork in a sample of secondary schools in England and Wales. *Review of Environmental Education Developments*, **14**, 11-13.

Gayford, C.G. (1986b). Environmental education and the secondary school curriculum. *Journal of Curriculum Studies*, **18**, 147-157.

Kelly, P.J. and Nicodemus, R.B. (1973). Early stages in the diffusion of the Nuffield A-Level biological science project. *Journal of Biological Education*. **7**, 15-22.

Kerr, J.F. (1963). *Practical Work in School Science*. Leicester: Leicester University Press.

Leal Filho, W.D.S. (1987). *Ecologia e Educacao*. Salvador: SEEB.

Leal Filho, W.D.S. (1991). *Strategies for the Implementation and Development of Environmental Education in Brazil*. Salvador: ABEA.

Masterton, T.H. (1973). *Environmental Studies: A Concentric Approach*. Edinburgh: Oliver & Boyd.

Mazzolenis, S. (1989). *Almanaque Abril*. Rio de Janeiro: Editora Abril.

Morgan, E. (1982). The establishment of sequential development of course aims for fieldwork in ecology and degree and diploma courses. In *Environmental Science: Teaching and Practice*, pp. 41.45. Proceedings of the Environmental Studies and Sciences in Higher Education. Northallerton: Emjock Press.

Morine, J. (1983). *L'Espace Naturel*. Paris: Editions du Seuil.

National Association of Field Studies Officers (NAFSO) (1989). *The Value of Field Studies Centres*. Cornwall, UK: NAFSO.

National Association of Outdoor Education (1972). *Outdoor Education*. High Melton, Doncaster: NAOE.

Nicholson, E.M. (1968). *Handbook of the Conservation Section of the International Biological Programme*. Oxford: Blackwell Scientific.

Parker, T.M. and Meldrum, K.I. (1973). *Outdoor Education*. London: J.M. Dent.

Parry, M. (1987). Planning and implementing environmental curriculum initiatives in primary and secondary schools in England and Wales. *National Association for Environmental Education*, occasional paper 11.

Perring, F. (1969). The demand for and supply of field biologists. *Journal of Biological Education*, **3**, 123-129.

Prosser, P. (1982). *The World in Your Doorstep: the Teacher, the Environment and Integrated Studies*. Plymouth: McGraw-Hill.

Schools Council Geography Committee (SCGC). (1980). *Outdoor Education in Secondary Schools*. Manchester: SCGC.

Selmes, C. (1973). Nuffield A-Level biology: attitudes to Science. *Journal of Biological Education*, **7**, 19-24.

Shepley, A.V. and Coultas, B.A. (1971). A field study of animal behaviour. *Journal of Biological Education*, **5**, 76-79.

Sinker, P.A. (1979). The aims of field work. *Review of Environmental Education Developments*, **7**, 6-11.

Study Group on Education and Field Biology (SGEFB) (1963). *Science Out of Doors*. London: Longman.

Webster, P.L. (1981). The Environment and Education: a Study of Policies of Selected Educational Institutions in South Wales. MEd. thesis, University of Wales, Cardiff.

# 8

# Learning about ecology through contact with vegetation

MARGARETE R. HARVEY

*Landscape Architect Milwaukee, WI, USA*

## Introduction

The call to inculcate an environmental ethic in our children (Seymour and Girardet, 1987), a land ethic (Leopold, 1949; Meine, 1987), or an outdoor ethic (Report of the President's Commission, 1987) is universal. Yet there is no one clear-cut strategy to achieve that end.

The formal transmission of such an environmental ethic to future generations is usually assigned to environmental education in our school systems. Schools are charged with the transmission of established knowledge and the shaping of minds capable of drawing inferences and prioritizing values, i.e. having the basis of an environmental value system.

But education does not happen in a vacuum; there may be other, informal and complementary roads to instill environmental ethics in children. These involve the child's home and play environment, or 'situational factors' as Hines, Hungerford and Tomera (1986/7) propose in their 'model of responsible environmental behavior'. In their paper, 'The development of children's concern for the environment', Hart and Chawla (1981) suggest direct personal contact with natural settings as a way of stimulating concern, apart from the indirect and second-hand information gained through formal education. After carefully tracing children's development as a process of original confusion between the self and the world, followed by increasing differentiation, toward an ideal goal of a mature sense of interrelatedness of self and world, they hypothesize that 'biological experiences form a most important basis for the development of an environmental ethic' (1981, p. 278).

It is the premise of this study that the developmental process of the child can be influenced by characteristics of the physical settings, in which it grows up, without claiming it to be the major influence (David and Weinstein, 1987). The specific attributes of the physical setting under

investigation are the natural, vegetative elements around children's home
and play areas and on their school grounds.

## Literature review

Direct experiences with the natural environment include contact with
animals and plants. Tuan (1978) contends that human beings identify with
animals more closely than with any other aspect of nature. Isaacs (1930)
has concluded that children 'are on the whole more spontaneously
interested in animals than in plants' which tend to be seen as 'little more
than gifts and decorations' (p. 170).

Empirical work regarding biological experiences with animals and
children's reactions to them has begun in earnest (Kellert, 1983, 1985), but
is still rather limited with regard to vegetation. This paper is intended to
redress the balance to some degree by concentrating on vegetation.

The physical environment of school children has been researched extensively (Gump, 1978; Kurtz, 1978; Rivlin and Weinstein, 1984; Gump,
1987). Almost exclusively it is equated with the indoor school environment (for reviews see George, 1975; Weinstein, 1979). By contrast, the
outdoor school environment in which children wait before and after their
lessons, play during breaks and lunch hours, and spend their physical
education classes has rarely been evaluated from the developmental
standpoint.

To consider this open space an educational resource or outdoor laboratory for biology and environmental education is the logical extension of
John Dewey's call for an 'experience curriculum' in education (1947).
Such a use has been demanded in England and Germany since the 1960s
(Countryside Commission, 1965; Schools Council, 1974; Winkel, 1985).
Today the number of schools actively engaged in establishing natural
areas on their school grounds is growing, but there has not been a
systematic attempt to measure their impact on children's knowledge and
attitudes compared to less well endowed school yards.

The potential to shape children's awareness and response to their
environment through this daily contact has also been recognized by
R. Moore in his 'Project WEY', the Washington Environmental Yard in
Berkeley (1974, 1978, 1986). WEY was designed to accommodate simul-
taneously the play needs of children, the recreational and aesthetic needs
of the community, and the curricular needs of education.

Children's experiences with nature in two related environments have
received attention: playgrounds and outdoor education at camps, field
centers or other sites away from school.

Numerous observational studies evaluating *playgrounds* concluded that having facets such as children's degree of control over their environment (Bengtson, 1973), complexity of layout (Schneekloth, 1976), and manipulability (Nicholson, 1971; van Ryzin, 1978) is preferable from a developmental point of view. Most of these desirable traits on playgrounds could be achieved by imaginative use of vegetation. But typically vegetation as an attribute was not considered, nor was a specific developmental outcome measured. Also, it was not the school playground, but neighborhood playgrounds, which were subject to investigation, with the emphasis on the recreational rather than educational aspect. For instance, of the numerous, creative playgrounds illustrated in Rouard and Simon's *Children's Play Spaces* (1977) only four are in school yards.

Reviews of research on *outdoor educational experiences* conclude that the out-of-doors provides a more stimulating learning environment for relevant fields of study, if the outdoor education experience is of sufficient duration (Crompton and Sellar, 1981; Backman and Crompton, 1984). After an experimental study comparing indoor with outdoor environmental education, Howie (1974) recommended a combination of classroom preparation with outdoor experiences. Similarly, a direct comparison of teaching environmental education in a 'classroom only' setting and a 'combined classroom plus practical application' in Israel found the combined method more useful to the student (Blum, 1982).

In a nationally representative sample of American science teachers, Keown (1986) noted that 16% never used the outdoors and the majority of classes use outdoor resources fewer than three times a year. Factors restricting the use of the outdoors as a teaching resource were in order of importance: financing the travel, class size too large, lack of support from administration and few local sites of interest. The school landscape, if developed as a teaching resource, would alleviate most of these difficulties.

Reviews of studies of environmental concern and its correlates have found inconsistent patterns. Van Liere and Dunlap (1980) found negative associations between expressions of concern and age; they suggest that these positive attitudes among the younger generation may partially reflect the addition of environmental education to the curriculum. By way of confirmation, Jaus (1984) found that just two hours of education with regard to conservation, pollution and recycling resulted in significantly more protective attitudes to nature in the treatment group as opposed to the non-treatment group. Similar results had been obtained by Gifford, Hay and Boros (1982). Despite frequent speculation of positive effects of direct involvement with nature over a longer period (Horvat and Voelker,

1976; Perdue and Warder, 1981) there is a scarcity of factual information about any antecedents to environmental concern.

Since all these aspects, the formal education at school and the direct contact with nature at home and on school grounds, were suspected potential antecedents, the schematic model of hypothesized interrelations shown in Figure 8.1 was developed for this study. Not all of the variables measured can be discussed here.

## Methods

A survey of 995 school age children from 21 junior schools, from a continuum of inner city to rural locations in the south of England, was conducted in October 1986. Multiple methods were employed. Physical measurements of school grounds were combined with a survey of teachers and students.

The participating schools were selected on the recommendation of eight different county planning departments with the aim of maximum between-setting variance with regard to vegetation on school grounds. Extreme cases, i.e. schools with a great deal of vegetation and schools with very little vegetation, were overrepresented. To avoid a biased sample with 'leafy' middle class schools and 'asphalt jungle type' schools in working class areas, middle class schools with little vegetation and working class schools with a lot of vegetation were selected on purpose.

At each school one class of 8–9-year-olds and a second class of 10–11-year-old children were interviewed in October 1986. A question-naire was administered by systematic group interview of about one hour's duration to 995 8–11-year-old students. It contained a mixture of open-ended and multiple-choice botany questions, a cognitive map of the children's school landscape, two standardized attitude scales, questions asking for demographic information and an inventory of past experiences with vegetation. Of the students' questionnaires, 15.1% were incomplete. The results are based on the analysis of the remaining 845 completed questionnaires.

A short questionnaire for teachers established school practices regarding the use of the school landscape and time spent on formal instruction in the natural sciences.

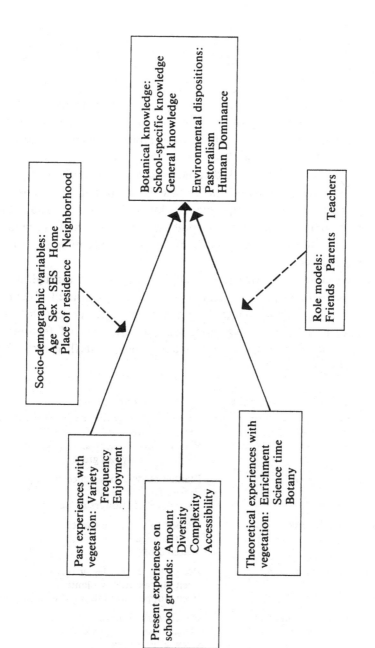

Figure 8.1. Model.

**Measuring instruments**

Children's *past experiences with vegetation* around their home and play areas were measured with regard to variety, frequency, and enjoyment (Harvey, 1989a).

A list of children's recollections and free associations regarding any contact with vegetation became the item pool for the measurement of past experiences. Items were grouped into six categories of past experiences with vegetation by content analysis; three typical items were selected to represent each category (see Table 8.1).

The questionnaire then asked subjects to indicate the *frequency* with which they had experienced these 18 activities (often, sometimes, or never) and the degree of *enjoyment* (a lot, a little, or not at all) they derived from these experiences. In addition, a simple count of the number of items a child had experienced furnished a measure of the *variety* of their experiences.

*Present experiences with vegetation* were established through actual physical measurements on the school grounds. They consisted of detailed landscape plans, based on Ordnance Survey maps and coupled with complete inventories of vegetation on the sites and all features of environmental relevance, such as wetlands, ponds, butterfly gardens, bird tables, greenhouses, etc.

Table 8.1. *Children's past experiences with vegetation options*

| | |
|---|---|
| Vegetation as a play object | Climbed a tree<br>Played hide and seek among bushes<br>Played in really tall grass |
| Vegetation as food | Planted seeds and watched them grow<br>Picked fruit or vegetables<br>Tasted leaves, flowers or berries |
| Vegetation as a task | Helped to weed the garden<br>Cut the lawn<br>Watered plants |
| Vegetation as an obstacle | Been stung by nettles or thistles<br>Been allergic to any plants<br>Been stopped from doing things by plants |
| Vegetation as an ornament | Put flowers in a vase<br>Pressed leaves or flowers<br>Grew a house plant |
| Vegetation as an adventure | Walked through a forest<br>Gone to play in a park<br>Gone camping in the countryside |

The plans were rank-ordered by six independent judges* for the *amount* and *diversity* of vegetation according to the plant inventories and the *complexity* of environmental features according to presence or absence of 16 additional features recorded on the site plans†. Information from teachers about school policy regarding access to the school grounds was used as a weight to arrive at a measure for *accessibility* of vegetation.

As for the dependent variables, *general knowledge of botany* was appraised through a mixture of 15 open-ended and multiple-choice botany questions from established achievement tests‡. School-specific knowledge of botany was calculated from a comparison of the cognitive map that each child drew of his or her school grounds and the official landscape plan.

*Environmental dispositions* were measured with two 22-item Likert-type scales from the Children's Environmental Response Inventory (Bunting and Cousins, 1983, 1985). They were 'Environmental Adaptation', here re-named '*Human Dominance*', because it purports to measure the belief in the right of human beings to use technology to adapt and dominate nature, and '*Pastoralism*' which expresses the appreciation of the natural environment in an intellectual and aesthetic fashion. A high score in Human Dominance is taken to show a lack of sympathy and identification between the self and the living environment and thus a deficient environmental ethic. A high score in Pastoralism is taken to indicate sympathy and identification with the environment.

In accordance with these definitions, Human Dominance was negatively related to Pastoralism ($r = -0.416$). General botanical knowledge was related positively to Pastoralism ($r = 0.387$) and negatively to Human Dominance ($r = -0.506$). The correlation coefficient for general and school-specific knowledge was 0.450.

## Results

The findings will be presented in four parts: 1. Some descriptive data concerning the children's past contact with vegetation; 2. An analysis of the

* Faculty members of the Department of Landscape Architecture at the University of Wisconsin-Madison.
† Items contributing to the complexity score were: Nature trails, meadows (long grass area with wild flowers), wetlands, ponds, plants in containers, bird tables and boxes, butterfly gardens, greenhouses, animals in cages, compost heaps, dead trees, hide-outs for wildlife observation, logpiles, tree nurseries, stiles, play structures.
‡ The items used were from MEAP (the Michigan Educational Assessment Program for grades 4 and 7), Science Form C; from NAEP (National Assessment of Educational Progress, the Third Assessment of Science, released excercise set, USA); and from APU (Assessment of Performance Unit, Department of Education and Science, London).

relationship between past contact and knowledge of botany and environmental dispositions; 3. The relation of children's present contact with
vegetation on school grounds and their knowledge of botany; and 4.
Analytical data regarding the interrelationships between all major variables under investigation.

The sample was evenly divided by sex: 425 girls and 420 boys. The ages
varied from 8 to 11 with the largest proportion at ages 10 (47.3%) and 9
(34.4%). Socio-economic status of students, measured by weighting
students' type of home with teacher's estimate of social class of school
catchment area, was normally distributed.

School size varied from 150 to 775 students (mean 327) with the potential area available for play per student ranging from 5.6 to 100.3 m$^2$
(mean 35.8 m$^2$). An equal number of schools were located in villages, the
rural/urban fringe and the inner city. Four schools were supported by the
Church of England, one was a Roman Catholic school. Five schools had
a large percentage of immigrant children.

The research project provided plentiful descriptive data regarding
children's past experiences with vegetation and important demographic
characteristics: sex, age, and socio-economic status. For a detailed
analysis, see Harvey (1989a).

### Gender differences

As a summary of recent empirical studies on gender differences concerning the environment, Chawla (1988) concluded that men have shown
greater knowledge of environmental facts, but women have shown
stronger environmental feelings and higher measures of concern. Could
this traditional sex-typing be based in part on differential experiences with
vegetation?

Table 8.2. *Past experience with vegetation by gender*

|                              | Girls |      | Boys  |      |
| ---------------------------- | ----- | ---- | ----- | ---- |
| Experience with vegetation   | Mean  | SD   | Mean  | SD   |
| Variety (1–18)               | 12.89 | 2.48 | 12.77 | 2.51 |
| Frequency (1–36)             | 19.10 | 5.21 | 18.74 | 5.18 |
| Enjoyment (1–36)             | 17.51 | 5.04 | 16.90 | 5.14 |
| N                            | (425) |      | (420) |      |

The findings with regard to gender are set out in Table 8.2. No sex differences can be registered when looking at the summary measures for variety of experiences, the frequency of experiences with vegetation, and the enjoyment thereof, i.e. boys experience vegetation as frequently as girls, their experiences are as varied, and they enjoy their contact just as much.

However, differences emerge when children's experiences with vegetation are grouped into the six categories described above (Table 8.3). Boys have significantly more contact with vegetation as an object of play and adventure, while girls encounter it significantly more as food and ornament and somewhat more as a task to be performed. This result mirrors R.C. Moore's observations that boys tended to use plant parts for pretend games of violence, while girls engaged more in small-scale play with plant parts (Moore, 1986). There was no sex difference with regard to vegetation being experienced as an obstacle.

The answers regarding the enjoyment of their experiences reveal a similar pattern: boys enjoyed contact with vegetation as play object and adventure significantly more than girls ($p = 0.000$). They also appreciated the challenge of vegetation as an obstacle. By contrast, girls liked their contact with vegetation as food and ornament more than boys.

### Age differences

The association between experiences with vegetation and age is straight forward: a steady accumulation of variety and frequency of experiences with age was to be expected and is indeed reflected by the means. But the increases are small, indicating that by age 8 to 11, new and additional experiences with vegetation (variety) are harder to come by and the

Table 8.3. *Past experience with vegetation (sub-categories) by gender*

| Experience with vegetation as: | Girls mean | Boys mean | $T$-statistic | $p$ |
|---|---|---|---|---|
| Play object (0–6) | 3.22 | 3.83 | 6.06 | 0.000 |
| Food (0–6) | 2.89 | 2.56 | 3.42 | 0.001 |
| Task (0–6) | 3.73 | 3.50 | 2.31 | 0.021 |
| Obstacle (0–6) | 2.45 | 2.54 | 1.08 | 0.280 |
| Ornament (0–6) | 3.44 | 2.44 | 9.37 | 0.000 |
| Adventure (0–6) | 3.39 | 3.90 | 5.42 | 0.000 |
| $N$ | (425) | (420) | | |

frequency of contact is also more or less established. Analysis of variance results for the four age groups and the summary measures of past experiences indicate that an age effect is present, but that it is weak. Measurements taken at younger ages may show a more dramatic age effect.

### Age and gender differences combined

Age and gender differences for the six categories of experience with vegetation are combined in Figure 8.2. The size of the gender differences are portrayed as the gaps between the male and female curves. It is noticeable that the gaps widen with age rather than converge. The one exception is experience with vegetation as an obstacle, where there are no differences. The consistently largest gap exists for using vegetation as an ornament.

The percentage of children reporting a low incidence of using vegetation as an object of play and adventure decreased steadily with age for males, while it increased for females at age 11. Contact with vegetation as a task changes little with age for either boys or girls, although girls have unfailingly more of it. As for using vegetation as an ornament, girls showed an overall decrease with age, whereas the boys' usage stayed more or less at a consistently low level. Finally, for vegetation as food the patterns for boys and girls (invariably more such contact) run parallel: increase in contact by age 9, decrease by age 10 and further increase by age 11.

### Status differences

Past studies have shown inconsistent relationships between social class or status and concern for the environment (Buttel and Flinn, 1978; Van Liere and Dunlap, 1980; Hines *et al.*, 1986/87). It was critical therefore to search for any pattern in the supposed antecedents to concern, the children's experiences with vegetation.

### Socio-economic differences

The data permitted the classification of all children according to their socio-economic status by combining the weightings for their type of home with the class composition of the school's neighborhood. A child living in a single family home in a middle class neighborhood would represent the highest status group (group 6), while a child living in an apartment in

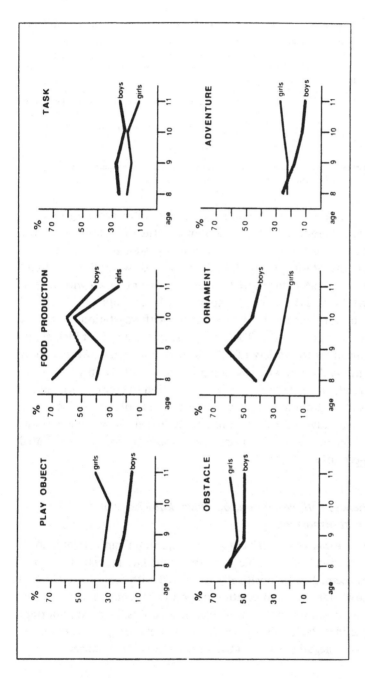

Figure 8.2. Percentage of children aged 8–11 who never or rarely experienced vegetation as one of six defined categories.

Table 8.4. *Past experiences with vegetation by socio-economic status: ANOVA*

| | Socio-economic status group (group means) | | | | | | | |
|---|---|---|---|---|---|---|---|---|
| Experience with vegetation | 1 | 2 | 3 | 4 | 5 | 6 | $F^*$ | $P$ |
| Variety | 11.8 | 12.1 | 12.8 | 12.9 | 13.5 | 14.5 | 12.18 | 0.000 |
| Frequency | 17.8 | 17.6 | 18.8 | 18.7 | 19.7 | 21.8 | 7.62 | 0.000 |
| Enjoyment | 16.3 | 15.9 | 17.6 | 17.0 | 17.9 | 19.5 | 5.59 | 0.000 |
| $N$ | 80 | 137 | 150 | 250 | 172 | 55 | | |

* Variance ratio distribution.

a working class area would be from the lowest status group (group 1).

Correlations between the socio-economic status of children and their past experiences with vegetation were low, but statistically significant. The higher the status group the more varied were the children's experiences ($r = 0.247$), the more frequent was their contact with vegetation ($r = 0.189$), and the more they appreciated this contact ($r = 0.159$).[†]

Analysis of variance results for experience with vegetation for the six status groups are presented in Table 8.4. The most consistent increase of the group means is for variety of experiences. Frequency of experiences increased gradually too, with the exception of status group 4.

Which kinds of experiences with vegetation increase with socio-economic status? Contact with vegetation as a play object, as food, as a task, and as an adventure were significantly higher, if the child's status group was higher. Only vegetation as an ornament was not associated with socio-economic status.

### Past experiences with vegetation, botanical knowledge and environmental dispositions

Students' *past experiences with vegetation* were significantly related to all four dependent variables (Table 8.5). In particular, the variety of past experiences seemed to be positively associated with school-specific and general knowledge. The appreciation of past experiences, i.e. the degree to which students enjoyed their past experiences, was the most important factor affecting Pastoralism. All three aspects of past experiences measured were negatively correlated with Human Dominance.

† For 843 degrees of freedom an $r = 0.089$ is significant at the $p < 0.01$ level of significance.

Table 8.5. *Pearson correlation matrix. Relations between past experiences with vegetation, botanical knowledge, and environmental dispositions*

| Past experiences | Botanical knowledge | | Environment disposition | |
|---|---|---|---|---|
| | General | School-specific | Human Dominance | Pastoralism |
| Variety | 0.308 | 0.266 | −0.283 | 0.304 |
| Frequency | 0.242 | 0.221 | −0.240 | 0.317 |
| Enjoyment | 0.193 | 0.157 | −0.197 | 0.359 |

$N = 794$, for Pastoralism $N = 388$.

*Note*: With a sample size of 845 (some infrequent responses not taken into account), small correlations will achieve statistical significance. To avoid too many significant, but relatively meaningless relationships, as well as circumventing the problem of changing sample size, the critical values for Pearson product–moment correlations for 100 degrees of freedom will be used, i.e. $r < 0.195$ is significant at the 0.05 probability level and $r < 0.254$ at the 0.01 level.

This means that by providing lots of varied opportunities for children to be in contact with plants, parents and teachers can contribute to the prevention of environmentally detrimental attitudes in their children to some extent. It can be a significant tool in the hands of concerned parents and teachers. Pastoralism, on the other hand, can be fostered through ensuring that children's contact with vegetation is positive, no matter how frequently it takes place.

There is not much differentiation when looking at the various sub-categories of experience. No matter in what context vegetation was encountered by the children, their tendency to see the environment as subordinate to the needs of humans was negatively impacted, while their love of the natural environment, Pastoralism, was positively correlated, even with such relatively unloved activities as food production and tasks.

Table 8.6 *Amount of vegetation on school grounds and general knowledge of botany (Row percentage)*

| Amount of vegetation | General knowledge of botany | | | | |
|---|---|---|---|---|---|
| | Low | – | Medium | – | High |
| Low | 20.2 | 28.3 | 26.2 | 13.7 | 11.6 |
| Medium | 13.5 | 28.9 | 34.2 | 15.1 | 8.2 |
| High | 7.6 | 20.3 | 29.4 | 24.6 | 18.0 |
| $N$ | 107 | 216 | 239 | 166 | 117 |

*Present contact with vegetation on school grounds*

The independent judges had grouped the 21 school yards into the wide categories of low, medium and high according to the amount and diversity of vegetation on the school grounds; each group contained seven schools. They had similarly designated the 21 school yards as low, medium or high in 'complexity' of environmental features. The teachers had rated the schools according to the ease with which students had 'access' to vegetation.

These environmental attributes were obviously not independent of each other. Amount of vegetation was highly correlated with diversity of vegetation ($r = 0.963$), complexity of environmental features ($r = 0.857$) and accessibility of vegetation ($r = 0.896$). Accordingly, the results for diversity, complexity, and accessibility of vegetation are very similar to 'amount of vegetation', so only one attribute will be presented here.

It had been hypothesized that students from green and complex school landscapes would score higher on botanical knowledge and Pastoralism and lower on Human Dominance.

This was borne out by the results. Table 8.6 shows a steady increase in the percentages of students' general botanical knowledge as the amount of vegetation on the school grounds goes up. More students from highly vegetated school grounds had high scores for general knowledge than students from schools with little vegetation.

The difference was more pronounced with regard to school-specific knowledge about vegetation (Table 8.7). Students who experienced a lot of vegetation in their school landscape included the correct amount of vegetation in their cognitive maps more frequently, they drew it at the approximate locations more often and they added the correct plant names more often (almost exclusively the names of trees).

Table 8.7. *Amount of vegetation on school grounds and school-specific knowledge of botany (Row percentage\*)*

|                       | School-specific knowledge of botany | | | | |
|-----------------------|------|------|--------|------|------|
| Amount of vegetation  | Low  | –    | Medium | –    | High |
| Low                   | 21.0 | 35.4 | 21.0   | 8.0  | 8.9  |
| Medium                | 16.1 | 16.5 | 24.0   | 20.4 | 14.5 |
| High                  | 7.5  | 12.0 | 31.0   | 23.2 | 24.3 |
| *N*                   | 115  | 170  | 225    | 154  | 152  |

\* Percentages do not add up to 100, because 29 of the 845 cognitive maps could not be reliably coded.

Table 8.8. *Complexity of environmental features on school grounds and students' environmental dispositions*

|  | Human Dominance | | Pastoralism | |
| --- | --- | --- | --- | --- |
| Complexity | Mean | SD | Mean | SD |
| Low | 60.33 | 8.40 | 81.68 | 12.11 |
| Medium | 57.83 | 10.03 | 87.31 | 10.44 |
| High | 54.28 | 10.27 | 90.01 | 9.96 |

The results for the *environmental dispositions* indicate that students from schools with vegetated landscapes tended to have higher scores for Pastoralism, which means that they agreed with statements like 'I really enjoy nature' or 'I like places where there are lots of plants and trees'. They also scored lower on the scale for Human Dominance over nature, which endorses statements like 'People should be able to cut down trees whenever they want to' or 'I am glad that people can change nature'. Table 8.8 gives the mean scale score for the groups of schools that the judges ranked low, medium, and high for complexity of environmental features. It shows a steady decrease in the Human Dominance scores as one progresses from the less complex to the complex school grounds, and conversely a steady increase in the Pastoralism scores.

To summarize the relationships between the four attributes of school grounds and the four dependent variables, Pearson correlation coefficients are presented in Table 8.9. All the correlations obtained are statistically significant at the $p < 0.01$ level of significance.

Table 8.9. *Pearson correlation matrix. Relations between environmental attributes of school grounds, botanical knowledge, and environmental disposition*

|  | Environmental attribute | | | |
| --- | --- | --- | --- | --- |
| Knowledge/dispositions | Amount | Diversity | Complexity | Access |
| School-specific knowledge | 0.373 | 0.357 | 0.367 | 0.369 |
| General knowledge | 0.215 | 0.224 | 0.293 | 0.215 |
| Pastoralism | 0.157 | 0.170 | 0.208 | 0.202 |
| Human Dominance | −0.232 | −0.250 | −0.266 | −0.212 |

Human Dominance $N = 796$, for Pastoralism $N = 388$.
For 843 degrees of freedom and $r = 0.089$ is significant at the $p < 0.01$ level of significance, while an $r = 0.068$ is significant at the $p < 0.05$ level.

Among the four dependent variables, school-specific knowledge of vegetation is the most highly correlated with attributes of the school grounds. For general botanical knowledge, the correlations are lower across the board, and for Pastoralism they are the lowest. As hypothesized, Human Dominance is negatively related to all four environmental attributes (Table 8.9).

Among the four natural attributes measured, complexity of environmental features achieved the highest correlations with general botanical knowledge, Pastoralism, and Human Dominance.

### Past and present contact with vegetation, environmental dispositions and botanical knowledge

As the data show both past and present experiences with vegetation to be associated with botanical knowledge and environmental dispositions, the question arises as to the relative importance of these factors in explaining the variance in the dependent variables. Accordingly, multiple regression analyses were undertaken with the addition of demographic factors (gender, age, socio-economic status) and the reported views of children's role models (parents, friends, teachers).

The analyses resulted in regression equations for the four dependent variables, which are summarized in Table 8.10. These four equations are attempts at multivariate models for the development of environmental dispositions and botanical knowledge. The multiple $R^2$ at the bottom of the table represents the percentage of the variance in the dependent variables explained by the models. The multiple $R^2$ range from 25.1% to 17.4%.

As far as the demographic factors were concerned, only age and the socio-economic status of students explained additional variance. General and specific botanical knowledge obviously improved somewhat with age, and children with higher socio-economic status had better general botanical knowledge, and scored higher on Pastoralism and lower on Human Dominance.

Among the reported role models, the views of friends were important for Pastoralism, those of parents for Human Dominance, but negatively so, and teachers counted in the development of school-specific knowledge of vegetation. Lastly, general knowledge depended to a small degree on 'enrichment', a category measuring the frequency with which the teachers provided instruction outside of the classroom.

Table 8.10. *Multiple regression analysis. Beta weights and multiple correlation coefficients for general and school-specific knowledge, Pastoralism, and Human Dominance*

| | Criterion variables | | | |
| --- | --- | --- | --- | --- |
| Predictor variables | Specific knowledge | General knowledge | Pastoralism | Human Dominance |
| *Present experience* | | | | |
| Amount | 0.376 | – | – | – |
| Complexity | – | 0.217 | 0.190 | –0.182 |
| *Past experience* | | | | |
| Variety | 0.186 | 0.249 | – | –0.215 |
| Appreciation | – | – | 0.320 | – |
| *Theoretical experience* | | | | |
| Enrichment | – | 0.076 | – | – |
| *Demographic* | | | | |
| Age | 0.242 | 0.197 | – | –0.085 |
| Socio-economic status | – | 0.119 | 0.127 | –0.147 |
| *Role models* | | | | |
| Friends | – | – | 0.175 | – |
| Parents | – | – | – | –0.114 |
| Teachers | 0.134 | – | – | – |
| *Multiple $R^2$* | 0.251 | 0.228 | 0.231 | 0.174 |

### Discussion and conclusion

In as much as this study provides evidence of a relationship between children's past and present contact with vegetation and their cognitive and affective development, it raises policy and design questions regarding the future planning of school landscapes, as well as all other children's outside play spaces.

As a caveat, statistical significance is not to be confused with real life importance. Before basing any action on these significant differences, it is wise to consider the size of the effect and the explained variance. The percentage of variance explained by the multiple regression models is low for all four dependent variables. But then no major influence had been expected.

The educational process is cumulative after all. No one factor, but a combination of physical, social and psychological factors, determine the efficacy of the educational process. It is further mediated by the personality of the child, the organismic specificity as Wachs (1987) calls it. None of the psychological variables could be included in this study; especially troubling is the lack of information on IQs or educational achievement. As only a fraction of these potentially relevant factors appear as variables in the model, it can be considered important that the variables included explain some of the variance at all.

Secondly, the importance of variables found to impinge on educational efficacy increases with the ease with which they can be manipulated. These variables do give us a tool to purposefully influence the developmental outcome through design.

All the environmental attributes of the school grounds measured for this study can be manipulated relatively easily. The one factor that seemed to account for most of the variance in the dependent variables was the *complexity of environmental features*. The more environmentally complex school grounds, not necessarily the most vegetated, offered the more effective learning experiences for the students in this sample as far as general botanical knowledge and the two environmental dispositions were concerned. This reinforces results obtained by developmental researchers who consistently report a positive relationship between variety and complexity of experiences and subsequent development.

This finding should be reassuring to schools with crowded and mostly hard top grounds without possibilities for expansion. Many of the features contributing to the complexity score in this study were small enough to be incorporated in cramped conditions, for example bird feeders, container plants, log piles, pets, compost bins, etc.

While diversity of vegetation did not contribute to explaining any additional variance, the amount of vegetation was associated with school-specific knowledge of vegetation. The reader may consider this self-evident. However, an equally plausible hypothesis could have claimed that vegetation as a relatively constant stimulus in the environment 'washes out' sensation and is subject to sensory adaptation, or that vegetation does not constitute the minimal perceptual change in an environment for it to be noticed. The results show that vegetation was relevant to the purpose and the activities of the subjects; they did indeed notice it.

Students' past experiences with vegetation are manipulable too. Parents and teachers can pursue the exposure of their charges to varied ecosystems, involve children in gardening and yard work, encourage them to decorate with plants and take them to parks, camps or the countryside.

The *variety of past experiences with vegetation* turned out to be the most important factor in the prevention of environmentally destructive dispositions such as Human Dominance. This means that the more varied the children's contact with vegetation had been, the less likely were they to believe that people could manipulate the environment with impunity.

This indicates that children's environments should be designed with as much variety of vegetation as possible, with some vegetation that lends itself to be used for play and adventure, other plant life that stands for food production and task fulfilment and yet other plants that function as obstacles or decoration only. Streets, playgrounds, parks or neighborhood nooks and crannies should provide a mix of vegetation rather than the usual large sweep of uniform vegetation such as mature trees on a manicured lawn or banks covered with low growing conifers lacking any seasonal changes.

The same applies to gardens of young families: while the demand is for landscapes with minimal or no maintenance, keeping a plot for food production and involving children in routine tasks within their garden, would be preferable for the development of positive environmental dispositions.

Pastoralism was more dependent on the *enjoyment* the children had derived from their experiences. Therefore, any vegetation present in the child's environment should have the potential to be enjoyed, i.e. it needs to be accessible and not a danger, nor should it have the potential to sting, scratch or poison the very young child. With increasing age 'dangerous' vegetation should be introduced under the heading of more variety. But apart from these intrinsic qualities, it is often the rules and regulations pertaining to the use of vegetation that determine whether children can enjoy it. Trees without any lower branches cannot be climbed; shrubs relegated to foundation plantings or trimmed to their bare bones make for uninspiring hide and seek games; expanses of lawn, while very conducive to running and ball games, hardly increase children's contact with nature.

While this is all relative (an inner city child may consider an expanse of lawn a luxury), there should obviously be room for both, the manicured vegetation supportive of other recreational activities and the raw, unpruned type, which demands attention in its own right and can withstand the onslaught of inquiring youngsters. The enjoyment of experiences is not so easily engineered. After all, experiences with vegetation as an obstacle or as a task, especially if repeated frequently, could have the opposite effect. The steady decrease in positive attitudes to vegetation with age reported elsewhere is a case in point (Harvey, 1989a, b).

Lastly, the question of relevance: So what, if botanical knowledge and environmental dispositions are slightly improved by the development of school landscapes? Botany is only a minor part of biology, and is frequently neglected in comparison to animal-related topics. In most American grade schools it is not part of the curriculum at all and in English junior schools it is more often a means rather than an end in itself.

The answer is twofold: First, it provides some empirical evidence in support of the quest to incorporate nature into the lives and play spaces of urban children (USDA Forest Service, 1977; Moore, G.T., 1985). Designing school landscapes with complex environmental features and diverse and accessible vegetation is not just an exercise in aesthetics; it is an educational resource that has largely been overlooked hitherto.

Secondly, the importance of the results hinges on the relation between botanical knowledge and environmental dispositions. The philosophy underlying the introduction of environmental education and its dissemination throughout the curriculum aims at the development of attitudes that help students acquire a set of values and feelings of concern for the environment as well as the motivation and commitment to participate in environmental maintenance and improvement (Engleson, 1985). A high score in Pastoralism and a low score in Human Dominance over nature can be considered operationalizations of the environmental ethics underlying environmental education. Any contributions toward the development of such attitudes, such developmental outcomes, are therefore to be welcomed.

Obviously, there must be many other variables impinging on the development of environmental dispositions. Consequently, the search for further variables, especially manipulable ones, needs to be iterative. The larger the number of predictors of environmental dispositions, the nearer we shall be to facilitating the learning of environmental ethics.

## References

Backman, S.J. and Crompton, J.L. (1984). Education experiences contribute to cognitive development. *Journal of Environmental Education*, **16**(2), 4-16.

Bengtson, A. (1973). *Adventure Playgrounds*. New York: Praeger.

Blum, A. (1982). Assessment of the subjective usefulness of an environmental science curriculum. *Science Education*, **66**(1), 25-34.

Bunting, T.E. and Cousins, L.R. (1983). Development and application of the 'Children's Environmental Response Inventory'. *Journal of Environmental Education*, **15**(1), 5-10.

Bunting, T.E. and Cousins, L.R. (1985) Environmental dispositions among school-age children. *Environment and Behavior*, **17**(6), 725-768.

Buttel, F.H. and Flinn, W.L. (1978). Social class and mass environmental beliefs, A reconsideration. *Environment and Behavior*, **10**(3), 433-450.

Chawla, L. (1988). *Children's Concern for the Natural Environment*. Paper presented at the American Psychological Conference, 1988.

Countryside Commission (1965). Conclusions and Recommendations of the Conference on Education (1965), *The Countryside in 1970, Second Conference*, University of Keele, March, 1965.

Crompton, J.L. and Sellar, C. (1981). Do outdoor education experiences contribute to positive development in the positive domain? *Journal of Environmental Education*, **12**(4), 21-29.

David, T.G. and Weinstein, C.S. (1987). The built environment and children's development. In *Spaces for Children*, ed. C.S. Weinstein and T.G. David, pp. 3-18. New York: Plenum.

Dewey, J. (1947). *Experience and Education*, 9th edn. New York: Macmillan.

Engleson, D.C. (1985). *A Guide to Curriculum Planning in Environmental Education*. Wisconsin Department of Public Instruction.

George, P.S. (1975). *Ten Years of Open Space Schools: A Review of the Research*. Gainesville: Florida Educational Research and Development Council.

Gifford, R., Hay, R., and Boros, K. (1982). Individual differences in environmental attitudes. *Journal of Environmental Education*, **14**(2), 19-23.

Gump, P.V. (1978). School Environments. In *Children and the Environment*. ed. I. Altman and J.F. Wohlwill, pp. 131-174. New York: Plenum.

Gump, P.V. (1987). School and classroom environments. In: *Handbook of Environmental Psychology*, Vol I, ed. D. Stokols and I. Altman, pp. 691-732. New York: Wiley.

Hart, R. and Chawla, L. (1981). The development of children's concern for the environment. *Zeitschrift fuer Umweltpolitik*, **4**, 271-294.

Harvey, M. R. (1989a). Children's experiences with vegetation. *Children's Environments Quarterly*, **6**(1), 36-43.

Harvey, M.R. (1989b). Children's experiences with vegetation on school grounds, their botanical knowledge and environmental dispositions. *Environmental Design Association*, Conference Proceedings, **20**, 256-263.

Hines, J.M., Hungerford, H.R. and Tomera, A.N. (1986/87). Analysis and synthesis of research on responsible environmental behavior: A meta-analysis. *Journal of Environmental Education*, **18**(2), 1-8.

Horvat, R.E. and Voelker, A.M. (1976). Using a Likert scale to measure 'Environmental Responsibility'. *Journal of Environmental Education*, **8**(1), 36-47.

Howie, T.R. (1974). Indoor and outdoor environmental education? *Journal of Environmental Education*, 6(2), 32–36.

Isaacs, S. (1930). *Intellectual Growth in Young Children*. New York: Harcourt, Brace.

Jaus, H.H. (1984) The development and retention of environmental attitudes in elementary school children. *Journal of Environmental Education*, 15(3), 33–36.

Kellert, S.R. (1983). Affective, cognitive, and evaluative perceptions of animals. In *Behavior and the Natural Environment*, ed. I. Altman and J. Wohlwill, pp. 241–267. New York: Plenum Press.

Kellert, S.R. (1985). Attitudes toward animals: Age-related development among children. *Journal of Environmental Education*, 16(3), 29–39.

Keown, Duane (1986). Teaching science in U.S. secondary schools: A survey. *Journal of Environmental Education*, 18(1), 23–29.

Kurtz, John M. (1978). Habitable schools: Programming for a changing environment. In *Facility Programming*. ed. W.F.E. Preiser. Stroudsburg, PA: Dowden, Hutchinson & Ross.

Leopold, Aldo (1949). *A Sand County Almanac*. Oxford University Press.

Meine, Curt (1987). Building 'The Land Ethic'. In *Companion to A Sand County Almanac*, ed. J. Baird Caldicott, pp. 172–185. Madison, WI: University of Wisconsin Press.

Moore, G.T. (1985). State of the art in play environment. In *When Children Play*. ed. J.L. Frost and S. Sunderlin, pp. 171–192. Wheaton, MD: Association for Childhood Education International.

Moore, Robin C. (1974). Anarchy zone: Encounters in a schoolyard. *Landscape Architecture*, Oct., 364–371.

Moore, Robin C. (1978). A WEY to design. *Journal of Architectural Education*, 31(4), 27–29.

Moore, Robin C. (1986). Nature's renewable play and learning resources: Plant parts as play props. *Playworld Journal*, 1(1), 3–6.

Nicholson, S. (1971). How not to cheat children: The theory of loose parts. *Landscape Architecture Magazine*, 62, 30–33.

Perdue, R.R. and Warder, D.S. (1981). Environmental education and attitude change. *Journal of Environmental Education*, 12(3), 25–28.

Report of the President's Commission (1987). *Americans Outdoors, The Legacy, the Challenge*. Washington, DC: Island Press.

Rivlin, L.G. and Weinstein, C.S. (1984). Education issues, school settings, and environmental psychology. *Journal of Environmental Psychology*, 4, 347–364.

Rouard, M. and Simon, J. (1977). *Children's Play Spaces*. New York: Overlook Press.

Schneekloth, L.H. (1976). *Environmental complexity and behavior of young children*. MSc.thesis, University of Wisconsin-Madison.

Schools Council (1974). *The School Outdoor Resource Area: Project Environment Schools Council*. London: Longmans.

Seymour, J. and Girardet, H. (1987). *Blueprint for a Green Planet*. London: Dorling Kindersley.

Tuan, Yi-Fu (1978). Children and the natural environment. In *Children and the Environment*, ed. I. Altman, and J. Wohlwill, pp. 1–32. New York: Plenum Press.

USDA Forest Service (1977). Children, Nature and the Urban Environment: Proceedings from a Symposium-Fair. *General Technical Report NE-30*, US Department of Agriculture, Forrest Service, PA.

Van Liere, K.D. and Dunlap, R.E. (1980). The social bases of environmental concern. *Public Opinion Quarterly*, **44**, 181–197.

van Ryzin, L. (1978). *Environmental manipulability in children's play settings.* PhD. thesis, University of Wisconsin-Madison.

Wachs, T.D. (1987). Developmental perspectives on designing for development. In *Spaces for Children*. ed. C.S. Weinstein, and T.G. David, New York: Plenum Press.

Weinstein, C.S. (1987). Designing preschool classrooms to support development. In *Spaces for Children*. ed. C.S. Weinstein, and T.G. David, New York: Plenum Press.

Weinstein, C.S. (1979). The physical environment of the school: A review of the research. *Review of Educational Research*, **49**(4), 577–610.

Winkel, G. (ed.) (1985). *Das Schulgarten Handbuch*. Seelze: Friedrich Verlag.

# 9

# Learning through landscapes

EILEEN ADAMS

*Formerly Director, Learning through Landscapes Project*

## Introduction

The *Learning through Landscapes Project* was a three year research project in the UK, from 1986 to 1990, to investigate the use, design, development and management of school grounds in primary and secondary schools. It was funded through a partnership between the Countryside Commission, a dozen local authorities and the Department of Education and Science (DES), The interim findings revealed that while it was generally agreed that the educational opportunities of school grounds should be enhanced and their quality improved, opinions differed as to how this might be achieved. The research helped to clarify the possibilities. The project report was published in 1990, addressed to those agencies who funded the research, but it is hoped that subsequent publications based on it, such as *The Outdoor Classroom*, published by the DES in 1990, will be helpful to teachers, headteachers, governors, designers and local authority officers in offering advice on how to extend educational opportunity and to improve environmental quality in school grounds.

In order to maximise the educational potential of the school environment, especially for ecological studies, it is necessary to ensure the landscape surrounding the school building provides a rich and varied resource for learning, whether this is an extensive estate or limited to a few planted areas. Even in inner city environments, ecological principles can be demonstrated and both long-term and short-term experiments established. Prior to the establishment of the Learning through Landscapes Project, little work had been carried out to examine the quality and potential of the school grounds as a resource for learning, with many initiatives being taken in an *ad hoc*, uncoordinated manner by highly motivated individual teachers and others.

The following account describes some of the findings of the research and raises some of the issues it reveals. This will include consideration of the design, development, management and use of school grounds and provide a glimpse of current work in the UK to create a more useful and stimulating environment for children to learn. Examples of the thinking behind the design of school grounds, of work in both primary and secondary schools and of initiatives to develop school grounds will be included.

**History of the use of school grounds**

When considering the design, development and management of school grounds, we must first clarify their functions. What are school grounds for? How do we use them? What purposes should they serve? The assumption has been that school grounds serve statutory requirements to provide for sports and games. Looking to the origins of the school landscape, we were aware of two separate elements: that of the play-ground and that of the sports field.

The provision for sports and games has its origin in the public schools (independent, not state-run) of the nineteenth century. The terminology was derived from field sports based on the hunting of game enjoyed by the upper classes. Games became a focus for 'character building' as well as for physical development. Physical education has continued to dominate the use of school grounds and influence their design throughout the century. As far as playgrounds are concerned, we find mention of a playground in a school as far back as 1821, 'equipped with circular swings, similar to maypoles, on which children could exercise'. A plan of a model school of 1863 shows a playground with swings, parallel bars and covered areas. The idea of providing for play was gradually dominated by the notion of exercise, first through formalised drill, then through broader physical education activities. The idea of play was subsumed by recreation. Official documents have focused mainly on physical play. Whilst the design of school buildings has changed dramatically to meet changing educational needs, the grounds seem to have changed very little from those of the nineteenth century school.

There is now, however, a move to extend the use of school grounds to encompass a wider range of educational activities. This has been heavily influenced by changes in learning and teaching methods. Much is known about children's need for richness and diversity of sensory experience, for varied encounters with the environment and the resulting stimulus and challenge it offers. School grounds provide an obvious and easily accessible environment to provide for this. They are a protected educational setting where opportunities to promote learning can be created in relation to the formal, the informal and the hidden curriculum.

## Use of school grounds

In primary schools, the growing emphasis on first-hand experience and investigative learning has encouraged the use of grounds as an extension of the classroom. Notions such as the outdoor laboratory and the sunshine classroom are increasingly popular. This is particularly evident in science studies linked with nature conservation and ecological awareness, where teachers are more prepared to use the landscape as a basis for study. A wealth of quantitive study goes on in school grounds, involving pupils in counting, measuring, surveying and estimating activities. Data banks are being developed with information on butterflies, bumble bees and the growth rates of potatoes. Studies of pond life, mapping exercises, archaeological digs, soil analyses, quadrat surveys, weather studies, experiments on growth, observational drawings of natural form, measurement of pitches and design studies concerned with changing the environment are all in a day's work in schools that value the resource that the grounds can offer.

But it is not only in teacher-directed study that the grounds are important. Providing for learning that is intrinsically motivated, that happens outside teachers' direction and that takes place outside lesson time is a significant function of the school grounds. What do pupils learn from their interactions in the school grounds during lesson breaks and lunchtimes, before and after school? Greater awareness of the educational value of play has led many schools to consider anew the environment the grounds provide for the informal curriculum. Those involved in ecology education will know of the importance of children having time to wonder, to explore, and to investigate, outside the constraints of lesson times. However, pupils learn not only from their teachers and other pupils, but from the environment itself. School grounds, just like school buildings, are important conveyors of messages and meanings, an external expression of the hidden curriculum. What messages about the relationships between people and place do young people receive from the normal school environment?

## Design of school grounds

School grounds can be used as a focus, a setting or a subject for study. They should be an outdoor learning environment that provides for learning and teaching activities to promote the physical, social and intellectual development of pupils within the context of the formal, informal and hidden curriculum. The grounds should offer a rich educational resource and accommodate learning and teaching styles that rely particularly on first-hand experience, investigative methods of study and independent learning. The grounds need to be seen as part of the continuum of learning experience inside and outside the school building, providing for a greater variety of practical, social and cultural activities and a wider range of stimulus than is possible within the confines of the classroom walls. Their design should be developed in answer to two fundamental questions: What kinds of experiences do we wish to provide for in school grounds? What learning and teaching activities will the grounds need to support?

The design of school grounds must take account of the context of the school, perhaps forming part of a 'nature corridor' of linked habitats for wildlife. It should provide a wide range of sensory experience and a variety of spaces for different activities. Microclimate should be considered in relation to provision for shelter, energy conservation and wildlife habitats. Provision needs to be made for access and circulation of pedestrians and vehicles. Care should be taken to include a variety of surface treatments and site furniture. The design should recognise the importance of landforms and soils to create interest in topography and planting. The soft landscaping should utilise a variety of planting, be robust and be capable of change and development. Safety and security should be prime considerations.

## Management of school grounds

However good their design and the educational thinking behind it, school grounds will not achieve their potential as an educational resource if the management is not sensitive and responsive. In the past, when the concern has primarily been the provision and maintenance of sports facilities, very little guidance has addressed the care and development of school grounds for a wider range of educational activities. The emphasis has been on grounds maintenance rather than on landscape management. Recent government initiatives have created new opportunities. As a result of the Education Reform Act of 1988 and associated legislation, schools will have greater control over their budgets, but also greater responsibility for managing their resources, which include the school grounds.

## Development of school grounds

Two separate lobbies have been influential on the development of school grounds: play and nature conservation. Greater ecological awareness and a concern for conserving plant and animal habitats has provided the impetus for many of the developments in school grounds. However, many current initiatives to develop school grounds are fragmented and of variable quality. Overwhelmed by the scale of the problem, many schools focus on one aspect, perhaps a play area, a nature conservation area, or more usually, a pond, and neglect its relationship with the whole school environment. An *ad hoc* approach to filling up the grounds with play structures, sculptures, planting and painted surfaces will not solve problems of scale, use of space or vandalism. Plans for school grounds development need to take account of curriculum policies and the changing needs of schools over a long period of time. The most successful developments have been those that have been linked with curriculum use. Different types of change are evident: physical changes to the school environment; changes in the behaviour of those who use it; and symbolic transformation of the environment to influence people's perceptions of it and their attitudes towards it.

None of the developments has been achieved without the schools overcoming many problems. Schools have lacked design expertise and do not necessarily know what will be involved in the development process. There are no easily accessible or readily available sources of advice and help. The systems or procedures are not in place to help schools in the UK. Developing school grounds involves both a design phase and a development phase: the design process is concerned with thinking about the nature of change and planning for it to happen; the development process is concerned with implementing the proposed change. It is important that schools learn as much as they can about these two phases so that they are fully prepared. We have identified a number of approaches: do-it-yourself; collaboration with an outside agency; and collaboration between school, outside agency and local authority. These are described in the following examples.

**Examples of school grounds developments**

The grounds of the Coombes Infant School in Berkshire UK have been developed over 20 years, with a bewildering yet planned diversity, as a rich and stimulating environment for young children to learn, a superb example of the do-it-yourself approach. The school policy is to make no distinction between the learning environment inside and outside the school building. Originally a flat site on the edge of a rural development, the grounds have been developed to provide contrast and variety of experience in all kinds of weather conditions throughout the different seasons and to encourage as much social interaction as possible. The soft landscape creates hundreds of opportunities for learning. Many different kinds of trees, fruit and vegetables are grown. There are ponds, mounds and ditches, hedgerows, compost heaps and protected habitats for small mammals. Sheep are kept on site and provision can be made to accommodate visiting animals. The hard landscape provides brick, concrete and wooden structures and painted markings which create endless opportunities for physical, social and imaginative play activities.

The grounds at Graham Balfour Comprehensive School in Staffordshire demonstrate what can be done over a period of 20 years through self-help. Here, the environmental science teacher has developed a rich and varied educational resource, creating school grounds of remarkable quality. Facilities include test beds for plant growing experiments, an arboretum of 64 species, a biology study area incorporating a pond and over 200 species of plants, animal pens, bee hives, tree nurseries, a coppice, flower beds, sculptures, greenhouses, storage sheds, workshops and compost heaps. The area is divided into a series of outdoor rooms, each with its own identity, yet part of a coherent whole.

The transformation of a playground at Weaverham County Primary School in Cheshire was a collaborative effort. Ideas for change were generated by the children and their teachers, an architect-parent developed the design, parents raised the necessary finance and a Manpower Services Commission team, funded by the local authority, did the construction work to create a quiet sheltered garden for children to work or relax and for parents to wait after school. Reading, writing, drawing, playing chess and tending plants are favourite occupations here. The garden is a place where children can become familiar with the many native trees, shrubs and flowers that form part of our literary and cultural heritage. The teachers felt that aesthetic education should not be forgotten in the excitement of scientific studies. In organising the planting, they considered it important to take account of the variation in shape and form of the plants, the range of leaf size and the association of colours, so that there were sufficient evergreen species to provide cover throughout the year and plants that would give a staggered flow of flowering, fruit and berry formation, foliage and bark colour.

Pupils and teachers at Gillespie Primary School in London have worked with an outside agency to develop their playground and improve opportunities for play. Through a number of design projects, artists from the Islington School Environmental Project have helped the pupils to develop a wider range of play activities. Although the focus has been on physical play, the introduction of structures and markings have encouraged different social encounters, greater cooperation between boys and girls and more opportunities for experiment in play activities. The structures have introduced more interesting forms and textures into the playground and have created important focal points for social and physical activity. Colour and pattern in the markings have created a visual link between the various parts of the playground.

The development of the grounds at West Walker Primary School in Newcastle upon Tyne has been the result of a partnership between the school, the Newcastle Architecture Workshop and the local education authority. It was a large scale project to provide for nature conservation and landscape enhancement as well as the provision of new facilities for more adventurous play. The overall scheme was designed to create an attractive and ecologically sound area to satisfy educational and recreational needs as well as extend natural habitats. There are a number of different zones, including: mounding to break the monotony of the site, create shelter for the school and echo the form of distant landscape views; shelter belt planting of native species to extend natural habitats and provide protection from wind and rain for the play areas and school buildings; a damp meadow planted with suitable wild species; a playground for school and community use; a nature garden for a field studies centre; seating and outdoor assembly area for visitors; and a hard surfaced area for the community cafe based in the school.

### Future development of school grounds

The time is right to reappraise the quality of the learning environment and the use we make of it. The *Learning through Landscapes* Report provides a framework to do this. It details the considerations that need to be taken into account to ensure a rich educational resource is made available for all pupils. Its recommendations cover the use, design, management and development of school grounds. They point to the need for the development of new attitudes that recognise the value of the school landscape as an educational resource. They call for convincing arguments to create a groundswell of interest and support for these. They advise that policies be established on the use, design and development of school grounds and new systems devised to improve the design, management and development of school grounds. They point to the need for the dissemination of information and advice on school grounds, particularly case study material, to identify and publicise organisations and individuals who can help schools, to provide an information exchange and to create support networks. They identify the need for further research and training and a development programme to promote good practice.

They reach out to policy makers and managers at all levels of government, including local school management, local authorities and central government, as well as groups such as teachers, headteachers, teacher educators and advisers. They are also directed at designers, parents and certain community agencies. They are intended to provide a focus for discussion and debate, to inform policy making and provide a framework for action. The next steps must be to formulate appropriate strategies and decide which agencies are best placed to put them into effect.

The recommendations assume a need for change in the use, design, management and development of school grounds and fall into various groups: attitudes; advocacy; policy; statutory provision; systems and procedures; roles and relationships; documentation and dissemination; information; advice; training; support networks; and a development programme. A Learning through Landscapes Trust has recently been set up to undertake a development programme. Its functions are to offer information and advice; to document and disseminate examples of good practice; and to provide advocacy on the use, design, development and management of school grounds. It will also promote further research; create an information exchange and a support network; and provide training programmes to ensure that all schools throughout the United Kingdom have access to a rich environment which will support a wide range of learning experiences.

**The challenge for schools**

The vision of the new school landscape is a challenging one. It rejects
our preconceptions of what the school environment is for and what it
should look like. It requires us to project into the future and to consider
the nature of schooling in the next century and consider what functions
the school environment might serve. It requires parents to imagine the
kind of place they would want their children to spend at least 11 years
of their lives; teachers to value experiential and investigatory learning
in school grounds, to reappraise the value of play in this context and
to consider how school grounds might constitute a resource for learning
and teaching; teacher educators to develop learning methods and teaching
strategies that exploit the resource the grounds can offer; headteachers
and governors to consider the best ways of managing this precious
resource, of caring for and developing it to create a rich and stimulating
setting for learning; designers to understand more about the processes
of learning and teaching and to confront anew the task of creating an
outdoor learning environment rich in aesthetic quality and educational
opportunity; community agencies to find ways of working that truly
permit user-participation in the design process; local authority administrators to devise policies and systems that support schools' efforts to
maintain and develop their grounds; central government to recognise
the importance of the school environment in children's educational
development and create the policy and statutory provision that creates
a firm basis for development. Above all, we need to understand how
to create a learning environment that both cherishes and challenges,
that promotes rather than inhibits learning and that can contribute in
some measure to schooldays being the 'happiest days of our life'.

The *Learning through Landscapes* Report is available from:
  The Learning through Landscapes Trust
  Third Floor, Southside Offices
  The Law Courts
  Winchester SO23 9DL
  UK.

# 10

# Ecology and environmental education in the urban environment

MONICA HALE

*Faculty of Human Sciences, London Guildhall University, Calcutta House, Old Castle Street, London E1 7NT, UK*

## Introduction

The changes to society brought about by industrial and agricultural development, the rise in urbanisation and the consumer society have resulted in a wave of social change in all developed and many developing nations. The effects on the environment and on society of the move towards urban-industrial systems has been the subject of a number of world reports such as *The Limits to Growth* (Meadows *et al.* 1956), *The World Conservation Strategy* (IUCN/UNEP/WWF, 1980), the Brundtland Report, *Our Common Future* (World Commission on Environment and Development, 1987), and *Caring for the Earth* (IUCN/UNEP/WWF, 1991).

These and other reports emphasise the need for conservation for survival and call for a greater environmental awareness of all people. Continuing environmental education, throughout life, is essential to this process of social change.

Concern regarding the way of life prompted by the city environment is not a recent phenomena, in the eighteenth century Rousseau (1750) drew attention to the 'paradox of the deterioration of mankind with material advances'. The role of ecology and its relationship to the deteriorating urban environment was recognised nearly 200 years later, when Elton (1926) observed that 'ecology as an area of zoology' was so important that it was 'more able to offer immediate practical help to mankind' than any of the other sciences; and 'in the present parlous state of civilisation, it would seem particularly important to include it in the training of young zoologists' (Hale, 1987).

Environmental problems can only be addressed by the identification, development and promotion of methods of environmental utilisation and interaction that are responsible, sustainable and careful. These must

be integrated into the way people carry out their profession, occupation
or trade. Environmental education is the key to educating children and
adults while, at the same time, enriching learning and supporting traditional educational objectives over a wide range of subjects and areas
of experience (Hale, 1992).

As knowledge about the environment has increased the necessity of
applying information from all natural and social science disciplines to
the solution of environmentally-related problems has been shown. A
global perspective on 'the environment' has become pervasive (Botkin,
1989).

This chapter examines the nature of urban ecology and assesses what
the urban environment has to offer environmental education. Areas of
action are proposed to improve the ecology and environmental quality of
towns and cities, and hence the resource for education in predominantly
urban areas. Examples of initiatives in environmental improvement and
enhancement of the educational resource that have taken place in London
over recent years are briefly described.

## Environmental education

Environmental education aims to raise people's awareness of their
environment by developing their knowledge and understanding of the
processes by which it is shaped. It should also aim to involve them in
environmental issues and the value judgments that need to be made at
a personal level. In order to achieve these aims people need to become
aware of environmental issues and acquire the background information
to enable them to make and influence decisions. Environmental education
is thus concerned with attitudes towards, and decisions about, environmental quality; with informed management of resources; and with the
ethical considerations that relate to these (National Curriculum Council,
1990).

Environmental education therefore needs to be incorporated into the
formal and non-formal sectors of education as well as in continuing
and adult education (see Thomas, Chapter 4).

The Resolution of the Council of the European Community and
Ministers of Education (in May 1988) stated that environmental education lays 'the foundations for a fully informed and active participation
of the individual in the protection of the environment and prudent and
rational use of natural resources' (EC Resolution 88/C177/03). This

Resolution urges that environmental education should be promoted as a matter of priority within all schools of the Community.

The significance of environmental education has been recognised in the UK Government's White Paper on the environment, *This Common Inheritance* (DoE, 1990). This devotes a chapter to 'Knowledge, education and training'. Environmental education is recognised as necessary to ensure the effectiveness of environmental policy through an aware and informed population (Hale, 1993).

The objectives of environmental education have been summarised by UNESCO (1977) as:

1. to foster a clear awareness of and concern about economic, social, political and ecological interdependence in urban and rural areas;
2. to provide every person with opportunities to acquire the knowledge, values, attitudes, commitment and skills needed to protect and improve the environment;
3. to create new patterns of behaviour of individuals, groups and society as a whole towards the environment.

All subjects in the curriculum can contribute in different ways to developing environmental awareness. This should not be seen as an additional burden for schools and institutions of higher and further education. Environmental education is about improving the quality of experience of the whole curriculum, using curriculum time more effectively, and the imaginative use of teaching approaches and resources (National Curriculum Council, 1990; see Hale and Hardie, Chapter 2).

Environmental education can enable people to gain experience of and become concerned for the environment in a number of contexts and appreciate how human activity, past and present, causes environmental change. A range of environments should be studied including the school; the neighbourhood; contrasting environments that pupils may have visited; and distant environments in their own country and the wider world.

Ecology underpins much of environmental education. As the majority of the population in developed countries live and work in the urban environment, opportunities need to be made to allow for the study of ecology in the local and wider environment.

The development of ecology areas within school grounds has been shown (Winkel, 1986) to enable people to:

- develop an emotional link to nature through personal experience;
- develop knowledge and understanding of the environment and its interconnections;
- evaluate such interconnections from individual, social, ecological or economic points of view;
- act (or not act) in the environment as a result of this evaluation.

Environmental training related to the work-place is as important as the environmental grounding received at school. While many large companies have developed sound environmental training for their employees, there is a lack of environmental training in medium to small size companies and in the service sector. Politicians, decision-makers and administrators in public authorities and government need basic environmental training as the achievement of improved environmental quality and sustainability is largely dependent upon their actions.

To achieve the level of environmental awareness and commitment necessary to ensure the environmental improvements outlined in the UK Government's and EC policies, adequate resources for environmental education and training in the urban environment should be provided. Opportunities and motivation towards caring for their environment need to be provided for both children and adults.

**Urban ecology**

Dawe and Kunz (1986) described urban ecology as the part of ecology that is specifically concerned with towns and cities, and observed:

Many of our urban problems are the result of the unwise management of resources and of a failure to see that environmental quality has an importance which cannot and should not be measured in instant economic terms. Urban ecology aims to look at the urban environment as a whole, and to discover and explain the underlying processes which determine urban conditions. Educationally, it has an important role to play in making people (in particular those who live in cities) aware of the ecological processes which affect them and which they themselves influence.

Urban ecology is now a recognised area of study defined as 'the study of biotic and abiotic components of urban ecosystems using ecological methods' (Sukopp and Hejny, 1991). In contrast to natural ecosystems, urban ecosystems are influenced by external social and economic factors. To achieve a theory of urban systems, studies of human factors are necessary. There is a considerable amount of work being carried out to achieve a theory of urban ecosystems. For example, of the 70 projects that have been carried out in the UNESCO Programme on Man and the Biosphere (MAB), 11 are on urban systems in different parts of the world.

Ecological aspects of urban ecosystems are now one of the four largest sections of the MAB programme.

Sukopp and Hejny (op. cit.) believe that it is the 'obviously inadequate adaptation of human societies to their surroundings which places ecology at the centre of the environmental discussion'. There is therefore a need for ecological understanding about the relationship between our society and its environment based on the knowledge of natural living conditions. Urban ecology has an important role to play as various forms of town, and cities as they have developed through history, represent important forms of human coexistence and relationships between people and their environment.

Ecologists were late to turn their attention to towns and cities and even today towns and nature are still an apparent contradiction for many people. Documentation from the middle of the nineteenth century describing changes in the flora and fauna of cities exists, although until recently few studies have determined the ecological relationships between plants and animals in the urban environment (Bornkamm *et al.*, 1982). It is often a surprise to many that in the built environment characteristic combinations of organisms can be found under similar conditions as distinct assemblages in different towns and cities.

Scientific insights into the conservation and development of nature and landscape in towns and cities need to be applied to landscape planning and education. Many of the problems of urban areas are a result of inadequate adaptation of society to its surroundings. Solutions can only be found if consideration is taken of the ecological requirements and social needs of society. Ecological criteria should serve to demonstrate the consequences of actions for the environment, for society and for the individual (Muller, 1978).

## The urban environment as an educational resource

It is important that learning should take place not only in the classroom, but also in the local and wider environment. First-hand experience of familiar and unfamiliar environments is an integral part of environmental education. Schools and higher education institutes need to consider the important contribution that can be made by outdoor education, both locally based and of longer duration further afield (National Curriculum Council, 1990; and Hale and Hardie, Chapter 2).

The educational value of ecological field-work has been widely recognised (Hale, 1986). The use of local sites in towns and cities, where over 80% of young adults in Britain go to school, promotes a sense of place as this field-work has direct relevance to the environment in which they live.

Ecological field-work advances the understanding of natural processes and their interactions. However, by itself the detailed description of different habitats and the identification of species does not fulfil this objective. The urban environment, even in the most built-up areas, provides the educational potential to demonstrate ecological principles and concepts, to extend investigations across other disciplines, and to develop a range of skills.

In fulfilling these aims, ecological parks and open spaces help to enrich the urban environment by providing a resource for wildlife, relaxation, recreation and ecological and environmental education. In Britain there is an increasing demand for urban nature areas as the benefits of this land use are becoming more apparent and general awareness of ecology and environmental quality continues to rise.

There have been a wide variety of approaches to urban ecology, from international campaigns to projects on a local scale. These include ecological investigations, education, recreation, environmental protection, town planning, the preservation of cultural relics, and sociological investigations (Barker, 1986).


**Nature conservation in cities**

In Europe the conservation of various species and the establishment of areas for wildlife and natural and semi-natural landscapes have been considered necessary for many years. However, it was not until recently that nature and landscape protection were also considered important in those areas where human influence is strongest and most enduring – in cities and metropolitan areas (Council of Europe, 1987).

Although urban areas account for less than 5% of the total land area in most European countries, more Europeans live in cities and metropolitan areas than in rural areas. There is still a heavy demand for new urban development which is still largely an unchecked problem. As well as the repercussions of urbanisation resulting from the direct demand for undeveloped land, many indirect consequences reach farther than the city's boundaries.

Nature areas in cities benefit wildlife and have a positive effect on the environment: larger open spaces help to cleanse the air through absorbing pollutants, improve water quality, contribute to the absorption of storm water, and have a cooling effect on air temperature (Johnston, 1990; Commission of the European Communities, 1990).

Over the past decade research into the plant and animal life of urban areas has revealed a surprising richness. Gradually, city planning authorities are becoming aware of the need for protection and preservation of wildlife within their boundaries. There have been a number of moves towards urban 'greening' in many European and British cities, which have met with varying degrees of success.

While considerable efforts have been made at local level to conserve nature and enrich the flora and fauna in the city by local communities, these actions are only ephemeral unless connected with the development of an overall environment protection programme.

Different attitudes to conservation exist in relation to cultural and national conditions. For example, in the United States and Britain, 'enjoyment' and 'education' are stressed in connection with proposed greening of urban areas; while in the Federal Republic of Germany, ethical reasons are prominent (Council of Europe, 1987).

Two basic concepts for nature conservation in city environments are: firstly, that the city should be merged into the surrounding landscape as far as possible; and secondly, that the alienation of the urban population from natural vegetation and animals should be eliminated. The close interweaving of habitats and natural areas should be a prime objective.

In densely built-up city centres from 80% to 100% of the surface per city block is paved. In providing for open space it is not yet clear whether it is better to have a few large open spaces and many small ones or to have just a limited number of large green surfaces. Provision of distributed green areas is most beneficial, being consistent with the basic objective of making closer contact with nature possible, particularly in view of the fact that more than 70% of people's free time is spent in an area directly around the home (Council of Europe, op. cit.).

In terms of how much 'nature' is necessary, consideration should be given to the energetics of ecosystems and their functional interrelation and the carrying capacity of the ecosystem. There is still a lack of studies that integrate analyses of specific ecological elements, such as climate, soil, or flora, in the urban environment (Bullock and Gregory, 1991). Most importantly, there is a need for a theory that satisfactorily explains the specific qualities of the urban industrial system from the ecological viewpoint (Pietsch, 1984).

The merging of the city into natural aspects requires the native flora and fauna to be encouraged. A diversity of areas and time scales are required for colonisation by particular plants and animals. The regions in which cities are located should be mirrored in their green areas. Development of diversity over time means that successional changes and the different stages of ecological succession be allowed to run their courses (Council of Europe, op. cit.).

For the development of existing species it is desirable that green areas in cities are connected to one another and to outer areas to form a 'biotope group system' (Deutscher Rat fur Landespflege, 1983). Attempts need to be made to build, replenish and develop food chains.

The succession of advanced ecological systems, especially those with a tendency to stability, requires particular protection and support. However, the city will always be a specific habitat with high dynamics. Protection of the resulting biotopes and biotic communities that are typical of the city presents a special task for nature conservation in the city.

**Nature conservation in the UK**

Traditionally nature conservation in Britain has been associated with the recreational, aesthetic and inspirational value of wildlife rather than with its scientific and educational value (Harrison *et al.*, 1986). However, renewed emphasis on the personal and social benefits of nature conservation, especially in the context of the urban area, is significant. Harrison *et al.* (op. cit.) noted that it comes at a time when the future viability of the economic basis of our cities is being severely tested, yet even at a time of economic uncertainty, nature conservation is seen as an essential component of urban life vital for personal and collective well-being.

Harrison *et al.* (op. cit.) argue that the urban area represents the most potent evidence of the human conquest of nature, and any attempt to incorporate nature into the town, by creating reserves, landscape gardens, green-belts, etc., often stands in direct opposition to the economic and utilitarian values upon which urban growth is founded. As Thomas (1984) observed 'what is useful and productive is likely to be ugly and distasteful'.

However, the establishment of several Urban Wildlife Groups and the growing momentum of the 'greening the city' campaign, in addition to the support given by English Nature (formerly the Nature Conservancy Council) to a number of conservation initiatives in urban areas, point to a new attitude to nature conservation in Britain's cities.

### Planning the urban ecological resource

In Britain there is intense demand for land, especially in urban areas. Even land not used for housing, industrial development or utilities is frequently subject to conflicting demands for use such as sports pitches, formal parks and gardens, or for the protection and enjoyment of wildlife. If adequate provision is to be made for nature conservation and environmental education it is necessary not only to assess the present extent of the wildlife habitat, but also to make effective plans on how it can be protected, enhanced or, in some cases, created.

UK Government policy as expressed in Department of the Environment Circular (108/77) *Nature Conservation and Planning* encourages local authorities to take full account of nature conservation factors both in formulating structure and local plans and in consideration of individual planning applications. It emphasises that areas of nature conservation interest are by no means confined to traditionally beautiful areas of countryside but occur in towns as well as the countryside (Greater London Council, 1985).

Local plans should have regard to the potential of sites in serving the recreational, educational and social needs of local communities. The enhancement of existing habitats and creation of new ones is particularly important where people have little contact with nature in their everyday lives because they are not already close to accessible wildlife sites. Local councils have been requested to take into consideration ecological factors when reviewing proposals for land development.

Positive planning for nature conservation in cities is relatively recent. Some local authorities have not only accepted that wildlife is a 'good' thing to have in the city but are positively promoting an ecological dimension in all land-use decisions. The former Greater London Council (GLC) pioneered this approach in establishing an 'Ecology Section' in the Planning Department in 1982 with a team of six ecologists providing a strategic approach to ecology and nature conservation in London. The main work of the unit was the creation and management of habitats for wildlife (Chandler, 1984). It survives as the London Ecology Unit since the demise of the GLC. Many other local authorities have since adopted a similar approach.

The planning of New Towns in green belt areas near to existing large conurbations in the UK has produced new ideas and procedures. The quality of the environment and the existence of open space was promoted as a 'selling point' to attract both industry and residents. For example, in Milton Keynes and Warrington ecologists were involved from the beginning in planning and development. As well as carrying out ecological assessments of existing habitats and proposing their conservation they established maintenance and management regimes and developed an environmental education programme for the community and visitors. This was to help people appreciate the newly created landscapes and habitats and to engender a respect and care for them.

A crucial element in nature conservation in both these new towns has been the involvement of the local population. Naturalist and environment groups, schools and volunteers from various community groups have all contributed to the work and publicity. Schools have 'adopted' certain areas to look after and have created a range of habitats as 'ecology areas' in their school grounds. In this context, nature conservation is not regarded as being primarily concerned with wild flowers and animals, but is more concerned with people and the enjoyment of nature (Parkin *et al.*, 1986).

Increasingly, local authorities are making provision for nature reserves when planning the urban environment. More areas are being designated official 'Local Nature Reserves' and many boroughs are establishing official ecology committees as part of their planning departments.

### Case study: nature in London

London is one of the largest capital cities in the world with a population of nearly seven million located within 600 square miles. The richness of London's wildlife is considerable and is only just coming to be recognised: over 2000 wild plant species grow within 20 miles of the centre of the city, occurring in a range of habitats. The pressures on the continued survival of wildlife are considerable as traditional forms of management of open habitats are no longer carried out, and environmental conditions such as pollution, disturbance, fragmentation of areas and intense human pressure are apparent (Hale, 1990).

Derelict and disused land in London accounts for an area of 23 square miles (Goode, 1986). Many of these sites have remained unused for a considerable length of time and range from former industrial sites to docklands, quarries and land which, due to its unsuitability, has never been built upon.

While the most sensitive species have long since been lost, other more resilient species have established themselves in new artificial habitats. Built features also provide a variety of habitats for wildlife: buildings are suitable roosting sites for a number of migratory and native birds and bats, and functional areas such as cemeteries, railway embankments and playing fields attract a number of species.

There have been considerable increases in the wildlife of London during the past century (Goode, 1986). Legislation aimed at protecting wildlife has been accompanied by a perceptible increase in tolerance and interest in nature by the general population.

Numerous local protection societies exist in London, many of which are remarkably powerful. There is much enthusiasm and interest in the environment resulting from people's needs in relation to the natural world. A series of planning inquiries affecting open space in London during the past decade has resulted in the conservation lobby winning some notable victories over proposed plans for development. This shows that a new philosophy is becoming accepted.

In a densely populated city such as London, sympathetic management of existing open space is often not enough to ensure that wildlife is a part of everyone's daily experience. In inner city areas where open space is at a premium one solution is to create new habitats (Chandler, 1984). Where there is little surviving local wildlife, as in many parts of London, this development has taken the form of ecological parks and community gardens to encourage the colonisation of species.

## Conclusion

Work on the relationship between people and the environment over the past ten years has shown that nature must develop in close relation to the local inhabitants and their customs. Sukopp *et al.* (1980) emphasised that organisms and biological communities should be conserved to allow people direct contact with the natural elements of their environment. Only such open spaces can 'lead to the experience of natural beauty which permits coexistence between a nature existing in its own right and people who are free to determine their own actions in this space' (Nohl, 1985).

Increasingly, resources are being channelled into purchasing land for informal recreation and wildlife and towards their upkeep and management. Further research is being conducted into urban ecology and the benefits to the individual of increased contact with nature in the built environment.

While significant improvements have been made in our cities over the past decade there is still a long way to go in planning a coherent strategy for creating greener and more livable cities, which fully integrate nature into the urban fabric. The importance of the urban resource for environmental education has been summed up by George and McKinley (1974) who stated that:

Institutional Education will not be complete until all aspects of human society and the remnants of non-human nature have made their contributions. Somehow, we have forgotten that the real classroom is the whole world and that we are all in it.

## References

Barker, G. (1986). The links between ecological science, local authority action and community involvement in planning and land management for nature conservation in British cities. *International Seminar on the Use, Handling and Management of Urban Green Areas.* Barcelona 21–24 April. MAB-UNESCO Comite Expagnol.

Bornkamm, R., Lee, J.A. and Seaward, M.R.D. (1982). *Urban Ecology: the Second European Symposium.* Oxford: Blackwell.

Botkin, D.B. (1989). *Changing the Global Environment: Perspectives on Human Involvement.* New York: Academic Press.

Bullock, P. and Gregory, P.J. (eds.) (1991). *Soils in the Urban Environment.* Oxford: Blackwell.

Chandler, J. (1984). Green London. *ECOS*, 5(4), 5–11.

Commission of the European Communities (1990). *Green Paper on the Urban Environment: Communication from the Commission to the Council and Parliament.* COM(90) 218 final, Brussels.

Council of Europe (1987). *Development of Flora and Fauna in Urban Areas.* Strasbourg: European Committee for the Conservation of Nature and Natural Resources.

Dawe, G. and Kunz, C. (1986). *Practising Ecology in the City.* London: BASSAC.

Department of the Environment (1990). *This Common Inheritance.* White Paper. London: HMSO.

Deutscher Rat fur Landespflege (1983). *Integrierter Gebietsschutz.* Bonn: Schriftenreihe d. Deutschen Rates f. Landespflege 41.

Elton, C.S. (1926). *Animal Ecology.* London: Sidgwick and Jackson.

European Commission Resolution (1988). *Resolution 88/C 177/03.* Brussels: EC.

George, C.G. and McKinley, D. (1974). *Urban Ecology.* New York: McGraw-Hill.

Goode, D.A, (1986). *Wild London.* London: Michael Joseph.

Greater London Council (1985). *Nature Conservation Guidelines for London. Ecology Handbook 3.* London: GLC.

Hale, M. (1986). Approaches to ecological teaching: the educational potential of the local environment. *Journal of Biological Education*, 20(3), 179–184.

Hale, M. (1987). Urban ecology: a problem of definition? *Journal of Biological Education*, 21(1), 14–16.

Hale, M. (1990). The use and provision of urban land for ecology field teaching: recent developments in North London. In *Land-Use Change: Proceedings of the International Geographical Union Asahikawa-Sapporo International Symposium*, ed. R.D. Hill, pp. 176–192. Hong Kong University Press.

Hale, M. (1992). Towards more effective support for environmental education in developing countries. Unpublished paper to the OECD Development Centre Symposium May 1992.

Hale, M. (1993). Educating for sustainability in developing countries: the need for environmental education support. *The International Journal of Environmental Education and Information*, Salford.

Harrison, C., Limb, M. and Burgess, J. (1986) Nature in the city: popular values for a living world. *Journal of Environment Management*, 25, 374–362.

International Union for the Conservation of Nature and Natural Resources (IUCN), with the United Nations Environment Programme (UNEP) and the World Wildlife Fund (WWF) (1980). *The World Conservation Strategy.* Gland, Switzerland.

International Union for the Conservation of Nature and Natural Resources (IUCN), with the United Nations Environment Programme (UNEP) and the World Wildlife Fund (WWF) (1991). *Caring for the Earth: A Strategy for Sustainable Living.* Gland, Switzerland.

Johnston, J. (1990). *Nature Areas for City People.* London: London Ecology Unit.

Meadows, D.H., Meadows, D.L., Randers, J. and Behrens, W.W. (1956). *The Limits to Growth.* London: Pan.

Muller, P. (1978). Okologische Informationen fur die Raum-und Stadtplanung. *Schr. R. Siedlungsverband Ruhrkohlenbezirk*, 61, 49–80.

National Curriculum Council (1990). *Curriculum Guidance No.7: Environmental Education.* York: NCC.

Nohl, W. (1985). Asthetik und Pflanzenverwendung im stadtischen Freiraum. *Natur u. Landschaft*, 60, 305–309.

Parkin, I., Collis, I. and Wood, A.A. (1986). *Naturschutzstrategie fur die Metropolis. Naturschutz in englischen Stadten.* Vortrag 14. Okologie-Forum, Hamburg 1984.

Pietsch, J. (1984). Nutzungs-und wirkungsorientierte Blastungsermittlungen auf okologischer Grundlage. *Natur u. Landschaft*, 59, 129–132.

Rousseau, J-J. (1750). Discours sur les sciences et les arts. In *The First and Second Discourses*, ed. R.D. Masters. New York: St. Martin's Press.

Sukopp, H., Blume, HP., Elvers, H. and Horbert, M. (1980). Beitrage zur Stadtokologie von Berlin (West). *Landschaftsentwickl. und Umweltforschung*, 3, 1–225.

Sukopp, H. and Hejny, S. (1991). *Urban Ecology: Plants and Plant*

*Communities in Urban Environments*. The Hague, The Netherlands: SPB
    Academic Publishing bv.
Thomas, K. (1984). *Man and the Natural World: Changing Attitudes in
    England, 1500–1800*. Harmondsworth, Middlesex: Penguin.
UNESCO (1977). *The International Workshop on Environmental Education:
    Belgrade, October 1975: Final Report*. Paris: UNESCO.
World Commission on Environment and Development (1987). *Our Common
    Future*. Oxford: Oxford University Press.
Winkel, G. (1986). Aufgaben des Schulbiologiezentrums Hannover fur
    Umwelterziehung. *Verhandl. Ges. f. Okologie*, **14**, 501–504.

# 11

# Ecological concepts as a basis for environmental education in Indonesia

MOHAMAD SOERJANI

*Director, Centre for Research of Human Resources and the Environment, and Chairman, Postgraduate Study in Environmental Science, University of Indonesia, Jakarta*

## Background

From the time of the 1972 Stockholm Conference on the Human Environment there has been a growing awareness of the urgent need to promote sound environmental management. In Indonesia in 1982 'The Basic Provisions for the Management of the Living Environment' was enacted by law (Act No. 4/1982). This was followed by Government Regulation No. 29 on Environmental Impact Analysis (EIA) in 1986. As a result, there has been an increasing understanding of how human beings are inter-related, inter-act and are inter-dependent with other biotic and non-biotic components of the life system. The living environment must sustain not only human but also non-human components.

Thus on the one hand humans are part of the biotic system but on the other, have an overall responsibility for the management of the environment. This requires an understanding of what the environment is and of the opportunities for as well as the limitations to sustaining the existence and promoting the welfare of all. Ecology is a basic discipline supporting environmental education. In Indonesia, ecology is applied and integrated into environmental science education together with economics, geography, technology and other aspects of social sciences. Essentially, environmental science is based on ecological understanding and how to apply this as a basis of a way of life. Environmental science is, therefore, understood as applied ecology.

With the present concern for global environmental issues, such as global warming, holes in the ozone layer and acid rain, there is a need for a global understanding and collaboration among countries and nations in research and education that will lead to actions in environmental science and management.

**Introduction**

Indonesia is a country rich in natural resources, but lacks sufficient qualified personnel to manage these resources in a sustainable way. It is also recognised that individuals have a right and obligation to manage the environment. There is a great diversity of types of ecosystem and a range of ethnic groups in Indonesia. Environmental management is decentralised so that it is the equal right and responsibility of all communities in all provincial areas and developmental sectors. Consequently, there is a need for trained and qualified personnel in environmental education in each of the 27 provinces and the 17 sectors of development.

**Indonesia and its resources**

According to Indian and Chinese literary sources, the islands comprising the Republic of Indonesia have been known since the third century BC. The islands have an abundance of natural resources and in 1988–89 a population of over 176 million people. Indonesia is now in the second half of its fifth Five Year Development Plan. While the country is still facing many challenges to the achievement of prosperity and equality among all its people, its rich and natural resources are gradually being managed and developed in a more sustainable way.

*Natural resources*

The implementation of an environmental strategy is a formidable task in a country so large and diverse. Indonesia is the largest archipelago in the world, consisting of five main islands and about 30 smaller archipelagos, totalling 13 667 islands and islets, of which 6000 are inhabited. Geographically, it is located at the crossroads of two oceans, forming a bridge between two continents, a situation that facilitates not only a unique tropical climate and fertile land covered by thick tropical rainforests and replenished by volcanic eruptions, but also a rich supply of resources and minerals with significant economic importance.

Indonesia's geographic situation results in a unique variety of biological resources. The tropical Indonesian rainforest represents 10% of the global tropical rainforest. It contains 325 100 species ($\pm$ 17%) out of the total 1 917 600 species of the world. Indonesian vegetation abounds in timber species, mainly of the dipterocarp (*meranti*) family, which is one of the main sources of timber, resins, vegetable fats and other phytochemicals used as raw materials for industries, cosmetics, medicines, spices and other products. The richness of the Indonesian flora results in the population being heavily dependent on these natural resources to support basic needs. Approximately 6000 species of plants are known to be used directly or indirectly by the local people.

Indonesia is widely recognised as an important transition area between the Asian and Australian continents, with the fauna showing a number of distinct changes across the region. This area is known as 'Wallacea', after one of the earliest scientists to describe the changes in species composition across this region (see Figure 11.1).

Among the rare species found in the area are the orangutan (*Pongo pygmaeus*), the Komodo dragon (*Varanus komodoensis*), which is the world's largest lizard, several forms of black Sulawesi macaque or '*monyet hitam*' (*Macaca nigra*), a small number of one-horned rhinoceroses or '*badak Jawa*' (*Rhinoceros sondaicus*) found only in the Ujung Kulon reserve in West Java. Other important fauna include the colourful fish found in Indonesian waters.

Indonesia is rich in minerals and energy resources, which form the basis of industrial development. The most important products for the Indonesian economy are oil and natural gas. The main producers of these resources are the provinces of East Kalimantan and South Sumatra. However, recent off-shore drilling has opened new production areas including Jakarta Bay, Natuna Islands in the South China Sea, Madura Island and the Timor Sea (shared with Australia). Sources of electricity are geothermal vents, hydro-electric resources, coal, and, in the future, most probably nuclear energy.

Among mineral products, tin is the most important commodity, with Indonesia serving as the second largest producer in the world. Other minerals found throughout the country include bauxite, nickel, ferronickel, copper, iron-sand concentrates, gold and silver.

### Human population

Indonesia's human resources are also abundant, but at present a serious lack of education and technical skills, as well as uneven population disibution, hinder their optimal contribution to national development. Indonesia is the fifth largest country in the world, with an estimated population of 176 million and a relatively high rate (1.9%) of population growth. The island of Java accounts for only 6.9% of the land area but has 60% of the total population (see Table 11.1).

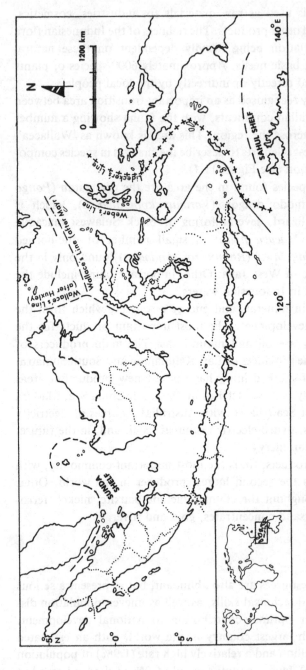

Figure 11.1. The Wallacea area of Indonesia. Wallacea includes the transitional zone between the Sunda and Sahul shelves. The western border of Wallacea is shown on the map as Wallace's line. Lidecker's line indicates the eastern boundary of the occurence of Oriental fauna, while Weber's line indicates the balance of 50% Australian and 50% Oriental fauna (Sandy, 1976, Soerjani, 1989a).

Table 11.1. *Estimated distribution of population (1988-1993)*

| Island(s) | Area (× 1000 km$^2$) | Total population (millions) | | Rate of growth (per year) (%) | Population density (per km$^2$) | |
|---|---|---|---|---|---|---|
| | | 1988 | 1993 (projected) | | 1988 | 1993 (projected) |
| Java and Kadura | 132.2 | 105.8 | 114.1 | 1.5 | 800 | 864 |
| Bali and Nusa Tenggara | 73.6 | 9.3 | 10.2 | 1.8 | 126 | 138 |
| Sumatera | 473.6 | 35.8 | 41.1 | 2.8 | 76 | 87 |
| Sulawesi | 189.2 | 12.3 | 13.3 | 1.7 | 65 | 71 |
| Kalimantan | 539.3 | 8.4 | 9.6 | 2.6 | 16 | 18 |
| Kaluku, Irian Jaya and Timor Timur | 511.3 | 4.0 | 4.5 | 2.6 | 8 | 9 |
| Indonesia | 1919.4 | 175.6 | 192.8 | 1.9 | 91 | 101 |

*Source:* Indonesian Government (1988-93).

The distribution of the population is becoming increasingly urbanised, a trend that is leading to urban environmental and social problems as well as a considerable decrease in the physical quality of the urban environment.

The labour force is also unevenly distributed. In 1986 there were 68.3 million workers, equivalent to 41.39% of the total population; of these, 42.2 million or 62% were located on the island of Java. Furthermore, nearly 50% of the workforce did not complete primary school education, while only 0.45% are university graduates (Figure 11.2).

**Environmental management and training**

Indonesia has formulated an environmental policy intended to incorporate the full participation of all people, where each person, both as an individual and as a community member, has an obligation to maintain the living environment and to prevent and abate environmental damage and pollution. One of the most important elements of this policy is Act No. 4 of 1982, 'The Basic Provisions for the Management of the Living Environment'.

The Government of Indonesia's present environmental concern is largely due to population pressure resulting from the high rate of popula-

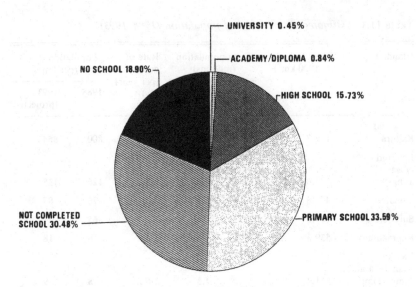

Figure 11.2. Educational attainment of the labour force (aged 10 years or over) in 1986. The high school labour force includes 6.60% junior high school, 3.69% secondary high school, and 5.44% vocational school (source: Central Bureau of Statistics, 1987).

tion growth and the inadequate level of education. This results in relatively low individual productivity. This situation has led most people to prioritise efforts to meet their basic needs at the cost of effort for proper environmental protection and management, thus leading to a deterioration in the quality of the environment. The solution for the improvement of the living conditions of the population is generally viewed as establishing and maintaining high rates of economic growth. However, it is also increasingly understood that development along the lines of a central economic growth, without considering the potential impact on the environment, will cause a deterioration in the quality of the physical, biological and social environment. The Environment Act is now accompanied by a requirement for environmental impact assessment (EIA). Under the present Government Regulation No. 29 (1986), which came in to effect in 1987, all development projects, both new and on-going, must take into consideration the possible impacts on the environment. Plans for environmental management and environmental monitoring procedures are proposed, based on the evaluation of the potential impacts.

However, with the existing condition and level of education in Indonesia, it is difficult to implement appropriate environmental manage-

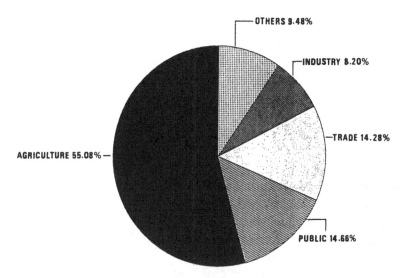

Figure 11.3. Distribution of workers in various sectors (source: Central Bureau of Statistics, 1987).

ment in the various sectors of development (Figure 11.2). To overcome this, the integration of environmental education into all disciplines at all levels is required. Environmental understanding is being implemented by integrating its principles and potential role with environmental skills across all disciplines so that environmental management is part of all training and education. Special training courses are organised such as in EIA (at basic and advanced levels), environmental administration, pollution prevention and management, population and environment, environmental law and environmental health. The distribution of workers in various sectors (Figure 11.3) gives an indication of the need to provide environmental training for workers in these sectors.

At present, degree programmes on the environment are only available at post-graduate level, leading to a Masters' degree. There are a limited number of doctorates in environmental science awarded each year. There is an urgent need to provide graduates in various disciplines with a basic understanding and relevant skills in a range of environmental areas. Most post-graduates in environmental science are those who are employed in various sectors of development, for example, in industry, trade, development consultancies, teaching, government, the military (and police) and non-governmental organisations (NGOs).

Environmental science is an applied form of ecology. It has a body of knowledge derived from ecological concepts, theories, principles,

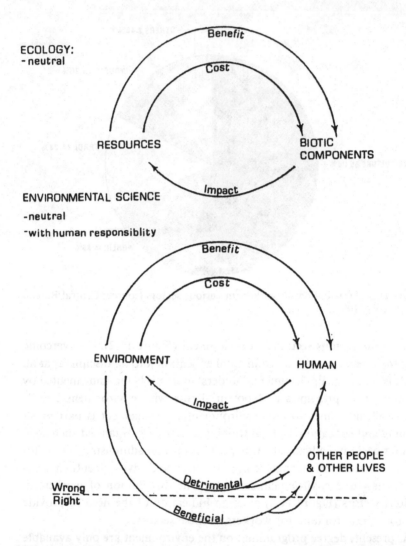

Figure 11.4. The comparison between ecology and environmental science, which indicates that environmental science is applied ecology (modifed from Beale, 1980 and Soerjani, 1988).

and laws. Environmental science incorporates knowledge of the environment in its totality in relation to how humans should behave and act in balance with nature in a sustainable way. Figure 11.4 shows the similarity between an ecological system and an environmental science system. From the anthropocentric view-point, both require a neutral and objective approach. However, environmental science promotes human responsibility as, depending on a range of human actions, these create impacts on other individuals. Therefore, environmental science has a value system that leads to the differentiation of a right or wrong concept.

**Nine key environmental concepts**

Everything in the environment is inter-connected. No categorisation of the variety of life could ever be regarded as complete or be assigned priorities (UNESCO–UNEP, 1990). The following nine key environmental concepts (based on, and modified from, Donella Meadows in UNESCO–UNEP, 1990), will be discussed below.

*1. The Gaia hypothesis*

The Gaia ('Earth goddess') hypothesis explains how organisms, in particular, micro-organisms, have evolved with the physical environment to provide an intricate control system which maintains the Earth's conditions favourable for life (Lovelock 1982, pp. 24–26; Myers 1985). This has been widely accepted as a basis for environmental perception, environmental ethics and environmental actions. In environmental ethics, this is a strong basis for sentientism, vitalism and altruism and in considering the human right as part of nature's right (Soerjani 1988). Our life is dependent on other biotic and non-biotic components not only of our own planet, but also of the whole solar system and the universe. Recently, Kasting *et al.* (1988) suggested, in a complementary way, that when life began on Earth, the carbonate–silicate geochemical cycle involving plankton and other organisms used these ions to construct shells of calcium carbonate resulting in a decrease of atmospheric carbon dioxide followed by a decrease in the Earth's temperature (from previous conditions to the present temperature of 13°C), so that more diversified life began. Thus, we are all created from the soil of the Earth. This concept strengthens altruism ethics as a basis of the physical aspects of environmental management.

*2. Noosphere*

The noosphere concept (noos (G) = mind) proposed by Vernadsky (Odum 1983, p. 53), suggested that the world is dominated by the human mind, so that it almost replaces the biosphere concept. Humans can

change natural systems (see Section 3 below) through the advancement
of science and technology. However, 'we don't fully understand how the
world works, we don't even understand how much we don't understand'
(UNESCO–UNEP, 1990, p. 4). We understand how the structure and
function of an organism enables it to survive and reproduce in the
environment in which it lives, but our ability to perceive the environment
through our senses and to conceptualise the environment in our minds
is severely limited (Ricklefs, 1978, pp. 19–26). Because of this, there is
a 'crisis of intelligibility', so that many human activities are implemented
under increasing uncertainty, i.e., decisions are taken although outcomes
are uncertain (UNESCO–UNEP, 1990, p. 4). Thus, we are often surprised
with unexpected and unpredictable impacts of our actions. The development of EIA is essentially as a tool to predict the consequences of certain
actions on the environment. Decisions will then be taken as to whether
and how the proposed actions should be implemented.

## 3. Levels of being

Human life begins in natural ecosystems, the biosphere or ecosphere.
However, with the present scientific and technological capability, environments have been created by human action. This has been referred to
as the technosphere. There is also a third component, the social environment or sociosphere (Figure 11.5).

The technosphere and sociosphere have played an important role in
the Earth's evolution. However, a wiser basis of environmental management must consider the underlying principles and laws of biological,
chemical and physical processes. This will contribute to better, balanced
environmental management.

## 4. Neutrality of nature

The anthropocentric view of nature widens the gap between the known
and the unknown. Host–parasite inter-relationships are not completely
understood (Lawton, 1984, pp. 334–45; Talbot, 1978, pp. 307–21).
Predators are perceived as cruel, voracious and guilty animals, while the
role of this inter-relationship in the resilience of a system, in a check
and balance system, in the survival of existence is often overlooked.
In a recent study of environmental education in Asian countries, it was
observed that there still exist inappropriate approaches to natural
phenomena in biological text-books. For example, the leaf of a tree is

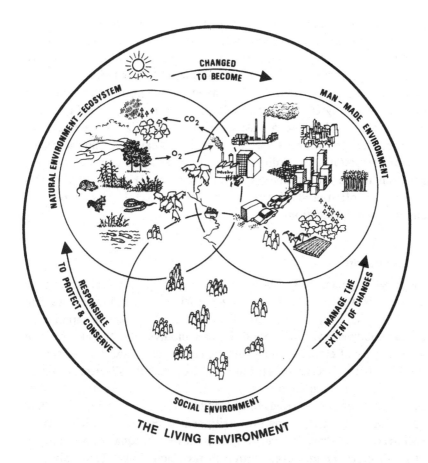

Figure 11.5. The concept of levels of being, in which balance and harmony between the three components of the environment is the objective of environmental management (Soerjani, 1988 and Soerjani, 1989a: 13).

mentioned as the 'kitchen', and photosynthesis is explained as 'preparations of foods', etc. (Soerjani, 1989b). In this respect, ecological textbooks promoting anthropocentric views (such as in Odum (1983) discussing wasted resources in agricultural systems) should be reviewed and amended. It is important to give consideration to this in environmental education programmes.

## 5. Levels of scoping

Physical, biological and social factors operate in a variety of ways, in different ecological areas, at various levels of complexity and on different time scales. For example, it takes about 100 years for a tropical rainforest to re-establish to a climax community. Approximately one million hectares of forest may be logged annually in Indonesia. While the productive forest might last for 50 years, it takes another 50 years before reforestation is successful (Soerjani, 1990). Therefore, environmental education must follow progressive contextualisation approaches (Vayda, 1982), in which certain phenomena must be analysed within a progressive range of space and time scales. Analyses of problems caused by an increasing rate of deforestation in Kalimantan need to take into consideration the situation in the timber market in Singapore and Hong Kong. In this respect the often-used phrase 'think globally, act locally' applies to environmental education.

## 6. Biodiversity

Biodiversity generally increases natural resilience. Natural ecosystems are organised into hierarchies that are finely tuned, stable and resilient (UNESCO–UNEP, 1990, p. 3). This concept can also be applied to human biodiversity. 'Horizontal' human biodiversity, the existence of a complexity of ethnic groups, cultural communities and jobs to fulfil a greater variety of niches, may promote environmental resilience. However, the 'vertical' human biodiversity, or social status diversity, i.e., gaps between rich and poor, needs to be bridged. Therefore, in environmental education, horizontal human diversity can be promoted, while vertical socio-economic diversity can only exist after an equal opportunity is given to everyone and every group in the community. Thus, diversity can only be tolerated if it is based on diversified responsibilities and achievements.

## 7. Cycles

The law of entropy reviews the fundamental concepts of matter and energy. This needs to be considered in relation to environmental education. The Earth is an open system as energy is dispersed into space and captured from the sun. Matter on Earth is generally considered to be locked in a closed system. However, in environmental management, reclamation of used resources such as iron, steel or other exhaustible and dispersed material is expensive or sometimes impossible. Therefore, environmental education must include all possibilities, such as reducing the consumption of resources, researching and developing substitutes, promoting the utilisation of renewable resources, producing durable products, etc. (see Miller, 1979, pp. 234–86).

## 8. Complex system

The most important components of environmental systems are: matter and energy, inter-connections and functions or purpose. Through their inter-connections, systems are more than the sum of their parts. In a hierarchial organisation, components are combined to produce larger functional wholes, and new properties may emerge that did not exist at the levels of its components. Accordingly, an emergent property of an ecological unit cannot be predicted from the study of the components of the level or unit (Odum, 1983, pp. 5–8). It has to be realised that the inter-connection (i.e. inter-relationship, inter-action and inter-dependence) of components of a system is organised into hierarchies. This means that the components of a system are not equally strongly inter-connected. In the application of this concept in environmental science education, it has to be realised that there are different levels of spatial distribution and time scales. The main issue in every environmental problem has to be identified as a basis for establishing priorities in its management. This is an essential approach or guideline already implemented in environmental impact analysis.

## 9. Sustainability

Sustainable development is built up on a number of concepts and definitions that need to be clarified before practical implementation is possible. If equitable human wealth is the objective of development, sustainable development must be carefully planned, implemented, managed and monitored with the support of natural resources and processes. The natural forces of planetary systems are vast compared to human forces. It is therefore easier to harness these forces than to work against them (UNESCO–UNEP, 1990, p. 3). Sustainable development is also understood as development within environmental parameters. Developmental projects must have not only technological appropriateness, but also environmental feasibility and socio-economic sustainability (Figure 11.6).

Socio-economic sustainability means that a project must be supported

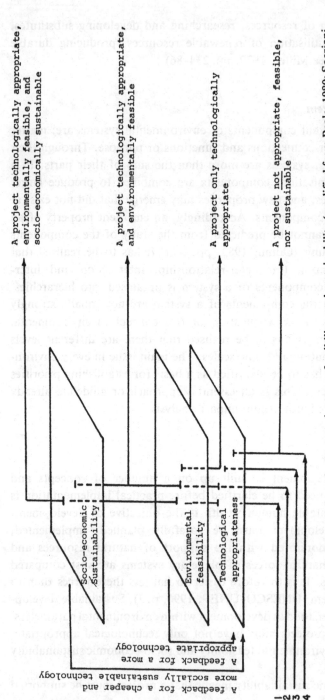

Figure 11.6. A process for screening projects for appropriateness, feasibility, and sustainability (modified from Beale, 1980; Soerjani, 1988, 1989a).

and a sense of ownership by the community promoted where the project is located, Ownership of a project does not necessarily mean that members of the community get a direct benefit or profit from it, but more importantly, that they have an equal opportunity to participate and to contribute to the project planning, construction and operational processes.

## Conclusion

It has become increasingly recognised that ultimately local problems can seriously contribute to global issues, for example, issues such as global warming, due to energy over-consumption, deforestation and other air pollution problems; holes in the protective ozone layer, due to the introduction of chemical elements into the upper atmosphere; acid rain; pollution of marine waters and atmosphere. These require a global understanding, solidarity and development of a network to facilitate collaboration. This can be initiated among universities (in research, teaching material and lecturer or student exchanges), among the non-governmental organisations (NGOs), and among professional societies. This may lead to more effective and beneficial activities in research, education and actions in environmental science and management. As in other developmental sectors, the 'bottom-up' expression of needs and wishes, plans and implementation in environmental management must be promoted and encouraged. There are examples in Indonesia as well as in other countries – public campaigns to develop a clean water supply (Jakarta), to clean irrigation canals (Sri Lanka), to plant commercial (fruit) trees on marginal land, the campaign of the Children Alliance for the Protection of the Environment (CAPE) to consume only environmentally clean products, are only a few examples of such action.

## References

Beale, J.G. (1980). *The Manager and the Environment: General Theory and Practice of Environmental Management.* Oxford: Pergamon Press.

Central Bureau of Statistics (1987). *Statistical Pocket Book of Indonesia 1986/1987.* Jakarta.

Indonesian Government (1988–93). *Fifth Five Year Development Plan.* Jakarta: Indonesian Government.

Kasting, J.P., Toon, O.B. and Pollack, J.B. (1988). How climate evolved in the terrestrial planets. *Scientific American,* **258**, 46–53.

Lawton, J.H. (1984). Herbivore community organisation: general models and specific tests with phytophagous insects. In ed. P.W. Price,

160                                                                    *Mohamad Soerjani*

C.N. Slobodchokoff and W.S. Gaud, pp. 344–345. *A New Ecology*, New
York: John Wiley & Sons.
Lovelock, J.E. (1982). *Gaia: A New Look of Life on Earth*. Oxford and New
York: Oxford University Press.
Miller, G.T. Jr., (1979). *Living in the Environment*, 2nd edn. Belmont,
California: Wadsworth.
Myers, N. (1985). *The Gaia Atlas of Planet Management*. London: Pan
Books.
Odum, E.P. (1983). *Basic Ecology*. Tokyo: Holt Saunders.
Sandy, I.M. (1976). *Atlas Indonesia* (In Indonesian). Third Book: *Natural
Environment/Man/Management of Resources*. Denpasar, Bali:
Dwidjendra.
Soerjani, M. (1988). *The Promotion of Environmental Science in the Effort to
Support Sustainable Development* (In Indonesian). Jakarta: University of
Indonesia.
Soerjani, M. (1989a). *Promoting Environmental Study Centres in Indonesia in
Support of Sustainable Development*. Jakarta: DESC–UNDP/World
Bank/GO1 – INS/82/009.
Soerjani, M. (1989b). *Environmental Education Programme in the Formal
School System and Teacher Training in Asian Countries*. Bangkok,
Thailand: UNEO.
Soerjani, M. (1990). Logging of tropical rain-forest and its problems (In
Indonesian) *Lecture*, Faculty of Post-Graduate Programme. Jakarta:
University of Indonesia.
Ricklefs, R.E. (1978). *Ecology* (10th edn.). MA: Chiron Press.
Talbot, L.M. (1978). The role of predators in ecosystem management. In ed.
M.W. Holgate and M.J. Woodman, pp. 307–321. *The Breakdown and
Restoration of Ecosystems*. New York: Plenum Press.
UNESCO–UNEP (1990). *Harvesting One Hundredfold. A Key Concept and
Case Studies in Environmental Education*. A condensed form of a paper
by Donella Meadows. UNESCO–UNEP.
Vayda, A.P. (1982). Progressive contextualisation: a method for integrated
social biological research in man and biosphere. (MAB) program.
*Proceedings of a Workshop on Ecological Bases for Rational Resource
Utilisation in the Humid Tropics*. Selangor, Malaysia: UPM.

# 12

# The present status of environmental education in Japan

MASAHITO YOSHIDA

*Research Officer, Nature Conservation Society of Japan (NACS-J),*
*2-8-1 Toranomon, Minato-ku, Tokyo 105, Japan*

## Introduction

The history of environmental education in Japan can be divided into four stages. The first stage was a nature education movement started by private school teachers in the 1910s. The second and third stages centred on conservation education and pollution education respectively. Currently, environmental education is being promoted, drawing on both of the previous stages.

This paper traces the development of ecological and environmental education in Japan through this evolution in order to assess its present position and status.

## Stage 1: Nature education

A few teachers, influenced by the Heimat-kunde movement in Europe and the nature study movement in the United States, campaigned for so-called 'nature education' for the younger age group at elementary school.

They held the first Science Education Seminar in 1919, in which Choken Moromizato, a teacher at a Seijo-gakuen, a private school, advocated the establishment of a new subject, termed 'nature education', to be taught in elementary schools. At that time science education in public schools was included in the curriculum of fifth grade (age 10–11) pupils and was devoid of any field studies. Moromizato considered it most important for younger pupils to experience nature directly with their own senses. In the proposed curriculum presented to the Science Education Seminar, there was an ideal model of a nature education programme for lower age groups. In this curriculum, pupils observed butterflies, caterpillars, cicadas and dragonflies in the field, took care of their school gardens, rice fields, and domestic animals and played with leaves or nuts, soap bubbles and magnifying glasses.

The teachers' efforts were rewarded in 1941 with the establishment of a new ordinance for primary education, but unfortunately the movement died out due to the commencement of the Second World War. After the war, the movement was revived in a small number of private schools such as the Seijo-gakuen where a new subject called 'strolling' was introduced. This replaced the teaching of science for the younger age groups. Generally, in most public schools, nature education for younger pupils followed science education but at a more basic level and still with a lack of practical field activities.

However, in April 1990, the Ministry of Education introduced, on a trial basis, a new subject for elementary schools called 'Life Environment Study'. The intention of this new subject is to allow younger pupils to observe their immediate surroundings. This includes the natural and social environment and will lead to an understanding of their relationship with the environment. In the first and second grade (ages 6–8), pupils generally feel a sympathy for plants and animals as they have more contact with them and take care of them.

With Japan's increasing economic growth, families have decreased in number of offspring and life and death situations are experienced more infrequently. There is general concern that pupils will lose their traditional respect for all forms of life. This change in perception was emphasised in a recent incident when a group of junior high school boys killed a tramp just for fun in Yokohama. This is unusual behaviour in a country such as Japan and was widely regarded with dismay. As a consequence the Ministry of Education has required that Life Environment Study to be taught in all schools in Japan from 1992.

**Stage 2: Conservation education**

Since the end of the Second World War there have been at least two main currents in environmental education. One is 'conservation education' and the other is 'pollution education'.

Influenced by advances in ecological theory, such as the concepts of 'food chains', 'pyramids of numbers', (Elton), 'biological community' (Clements and Schelford) and 'ecosystems' (Tansley), a nature observation movement began in the 1950s. This aimed at conservation education through the understanding of the balance of nature. Founded in 1955, the

Conservation Association of the Miura Peninsula (Kanagawa Prefecture) has been very active in promoting a nature observation movement. In their field excursions, students were prohibited from collecting any plants or insects. This was a new procedure as 'nature hikes' at that time were generally to identify plants and animals; this often resulted in the over-exploitation of plant and insect populations. The new methodology was to let pupils observe nature in its natural habitat.

In 1957 students of the Tokyo University of Education started a nature education camp for children. This was regularly held until the University was transferred to Tsukuba City.

In the same year the Nature Conservation Society of Japan (NACS-J) submitted 'The opinion paper on promoting conservation education' to the Ministry of Education and other relevant Ministries. In this report, NACS-J advocated that the Ministry of Education should develop conservation education to be included in compulsory education (elementary school and junior high school). NACS-J followed this up with an 'Opinion paper on promoting conservation education in higher education' submitted to the Ministries in 1960.

The rapid industrialisation of Japan during the current century resulted in many environmental problems becoming apparent all over the country in the 1970s. The forests suffered from road construction and other activities associated with logging; coastlines were damaged by landfill operations and shore protection works; and rivers and ponds were polluted by synthetic detergents. In 1972 a number of citizen groups gathered in Tokyo and established the Japan Union for Nature Conservation.

As a result of the work of the conservation movements, the importance of conservation education was recognised. A number of independent initiatives were started; citizens living along the Tama River started nature observation hikes to improve public awareness of river conservation and environmental groups who were concerned with the conservation of tidelands in Chiba Prefecture and Miyagi Prefecture initiated bird watching and tideland observation schemes. As a result of these movements, several tidelands were protected as wildlife refuges in the Tokyo Bay.

NACS-J has been involved in conservation education for several decades. The NACS-J held a conservation educators' workshop in 1978 to establish a nationwide network of conservation education leaders. The NACS-J now has 10 000 conservation educators registered from all over Japan.

## Stage 3: Pollution education

Teachers in public schools included pollution education in their curriculum in the 1970s as a result of concern about pollution-generated illnesses, such as Minamata diseases caused by water pollution, and the effects of photochemical oxidants caused by air pollution.

Teachers who worked around the worst-affected areas were obliged to attempt to protect their pupils from pollution. They developed teaching materials themselves as there were no textbooks on pollution education. One example of valuable pollution monitoring work was when a teacher who worked in Edogawa Ward, Tokyo found that a large number of frogs collected by his pupils at the Edo river were deformed. He traced the origin of the frogs with his students and as a result of the study they found that the Edo river was severely polluted by synthetic detergents. They postulated that this probably caused the deformation of the frogs.

In another area a school teacher found that photochemical oxidants cause spots on the leaves of morning glory (*pomoea purpurea, I. tricolor*). In an experiment his students grew morning glories both in school and in their home environment. The plants were used as biological indicators to monitor air pollution in their home town.

In 1970 the Diet (parliament) introduced the 'Basic Law for Environmental Pollution Control'. Since then the prefectural or city boards of education have encouraged teachers to provide teaching materials concerned with aspects of pollution. For example, Shiga Prefecture has suffered from pollution of Lake Biwako (the largest lake in Japan). This lake functioned as a source of water for the Kansai district and an ordinance to save the lake from eutrophication has been drawn up. To raise public awareness of pollution of Lake Biwako, the authorities promoted a clean lake campaign and organised a convention of environmental education in 1980.

## Stage 4: Environmental education

From the time of their introduction, conservation education and pollution education have been separately treated. Recently, however, actions of the Kiyosato Environmental Education Forum and the establishment of the Japan Environmental Education Academy have served to bring these two elements together.

The numbers of instructors engaged in environmental education increased during the 1980s. In addition to school teachers and citizen groups, environmental interpreters and environmental education leaders were required at visitor centres in the natural parks established in the 1980s. The first Kiyosato Environmental Education Forum held in 1987 provided an opportunity for environmental education specialists and others engaged in education to exchange their views on aspects of environmental education. These included environmental education programmes in elementary and junior high schools, programmes to attract younger participants, the role of volunteer leaders and coordinators in environmental education, the management and fund raising of environmental education organisations, rural development and environmental education. The Kiyosato Environmental Education Forum will continue for five years and will produce reports on current environmental education activities and assess future prospects for environmental education in Japan.

The Japan Environmental Education Academy was established in May 1990. At the foundation conference held at the Tokyo Gakugei University, 300 environmental education leaders gathered from all over the country. The participants of the conference included school teachers, professors of education divisions in universities, field conservation leaders employed by environmental pressure groups, national park and bird sanctuary interpreters and rangers. The Environmental Education Academy will publish an annual report promoting environmental education and will provide opportunities to exchange information about environmental education.

**Conclusion**

There is an urgent need to integrate the natural and the social sciences, ecology and economy, technology and philosophy, the conservation movement and the anti-pollution movement, and school education and social education, in order to solve the current global environmental problems. However, regrettably, each sector and activity has been separated by nearsighted leaders and sectionalism of the bureaucracy. Even in the process of preparation for the United Nations Conference on Environment and Development (UNCED) held in June 1992, difficulties arose in bringing together different areas of environmental interest.

In 1991 the Ministry of Education published the educational guideline for environmental education in secondary schools. That for primary schools was published in 1992. These publications will be epoch-making events in the history of environmental education in Japan, not only because they are the first environmental education materials published by the Ministry of Education, but because they were edited by a committee consisting of school teachers, board members of the Environmental Education Academy and an observer from the Environment Agency.

However, physical obstacles which inhibit cultural exchange between different sectors have to be overcome in order to enable these guidelines to be put in to practice. This must be the goal of environmental education.

# 13

# Ecological studies between schools in Europe and Scandinavia: the benefits to the curriculum of working across national boundaries

DAVID SHIRLEY

*David Shirley and Associates, 5 Trevor Road, Hitchin, Herts SG4 9TA, UK; formerly Conservation Officer, Education Department, Hertfordshire County Council, Hertford, UK*

## Introduction

The study of ecological and environmental phenomena at local and international levels is an increasingly important feature of syllabuses in many countries. Two case histories illustrate how schools in the United Kingdom have carried out collaborative investigations with students in Scandinavia and Europe.

In 1986, 180 schools in Hertfordshire (United Kingdom) and Norway jointly studied aspects of atmospheric pollution. Using simple, standardised equipment, they measured the pH of rainfall over a synchronised monitoring period and undertook a programme of ecological investigation. Results were exchanged and incorporated into national reports. Further stages of the project involved exchange visits for teachers and students, and investigations into the occurrence and effects of ozone at low altitudes.

In 1988, schools in the United Kingdom participated in the European Commission sponsored project 'Living indicators in Europe'. Ecological studies and recommendations for environmental improvements were produced for their local areas and incorporated into a national report.

An evaluation of both projects suggests that collaborative investigations present opportunities and challenges for teaching ecology and the appreciation of environmental problems.

## Background

In recent years, educationalists in Europe and Scandinavia have emphasised the need to raise the level of ecological and environmental awareness amongst young people (Greig *et al.*, 1987, 1989). In England and Wales this is reflected in the new National Curriculum (National Curriculum Council, 1988, 1990a) within the requirements for science and geography. Students aged from five to 16 may study ecology and make practical investigations within their own localities.

As learning progresses, students are required to extend their knowledge from the local to the national and international dimension. Opportunities for students to gain first-hand practical experience in the field are restricted by time, money and overall curriculum demands. Teachers and students have to rely on secondary sources: these often provide conflicting evidence and to the students they seem remote.

One way of reducing remoteness is for students in different countries to exchange and pool results and ideas arising from investigations within their own local areas. This involves linking schools through national and international networks. Such linking has been used for cultural and linguistic purposes for many years. However, advances in international communications have increased the possibilities of information exchange both in time and type.

The following two projects illustrate how the teaching of ecology through school links can be used to develop scientific training amongst young people and encourage an awareness of environmental issues from the local to the global. They also demonstrate how such work relates to curriculum requirements in different countries and the importance of a sound administrative protocol.

### The Hertfordshire–Norwegian schools link

Hertfordshire is an inland English county of 1630 square kilometres, situated to the north of London. It has a population of just under one million, several large towns with associated light industry, and large areas of farm land. The local education authority is responsible for 97 secondary and 475 primary schools, and has pioneered environmental education and ecological studies in the county over the past 25 years.

In geographical and administrative compactness, Hertfordshire contrasts strongly with its Scandinavian partner. The Norwegian Ministry of Education and Religion is responsible for schools serving a population of about five million. Within its national boundaries are extremes of landscape and climate that are not experienced in its English counterpart. However, like Hertfordshire it has a tradition of environmental and ecological teaching in its schools, which may be augmented by local and national conservation organisations.

The basis for establishing links between schools in Hertfordshire and Norway arose from the 1985 Oslo conference 'From Environment to Action'. Teachers and other educationalists from European and Scandinavian countries identified common problems encountered when teaching widespread aspects of ecological and environmental topics. The development of links between Norwegian and Hertfordshire schools was seen as a method of testing the value of collaborative studies across national boundaries. The links were administered and co-ordinated in Norway by Sissel Dobson, a teacher from Ungdomsskole, Kongsberg who also held an honorary post in the Norwegian Conservation Society. In Hertfordshire the administrative and co-ordinating functions were carried out by the author in his capacity as conservation officer and advisory teacher for the Hertfordshire Education Department's Advisory Service.

The initial focus of study for the project was the occurrence and effects of acid precipitation. This topic was selected because it was a common and controversial issue in the two countries and it had a place in the curriculum for students in most parts of the school age range.

The aims of the project were:

1. to carry out practical investigations into the occurrence and ecological consequences of acid precipitation;
2. to encourage links between schools in each country through which to exchange the experiences and results of their findings;
3. to pool the results from Norway and Hertfordshire with data from other sources to provide an international background to work at the local level.

All schools in Norway and Hertfordshire were invited to participate in the project and linked according to the age range of their classes. Initial recruitment of schools was achieved by publicity through the respective education departments and through newspapers and television. Over 90 pairs of schools were formed and each was sent background material with suggestions for project and experimental work to enable students to carry out ecological investigations appropriate to their ages. Further information, in printed, photographic transparency and video form, was also made available through The Royal Norwegian Embassy, the Norwegian Conservation Society, the Central Electricity Generating Board (UK) and television companies.

The first collaborative studies centred upon:

1. direct measurements of pH of rainfall over set periods;
2. investigations into terrestrial and aquatic systems.

**Direct measurements of pH of rainfall**

In 1985 the UK WATCH ran a pilot project to enable its members to measure the pH of rainwater by means of a simple collecting device and indicator strips. WATCH is a UK voluntary environmental organisation sponsored by a national newspaper, The Sunday Times, and The Royal Society for Nature Conservation. In 1986 WATCH established the 'Acid Drops Project' in conjunction with the Field Studies Council, British Petroleum and British Drug Houses (BDH). The project involved young people throughout the UK in measuring the pH of rainwater over a set time in their own localities. The co-ordinators of the Hertfordshire–Norwegian schools link negotiated with WATCH to use their testing kits and to pool results with the UK programme.

From 27th September to 24th October 1986 the linked schools in Norway and Hertfordshire carried out pH monitoring of rainfall. The apparatus consisted of a plastic container fixed to a pole at 1.5 m above ground level. Daily, a clean polythene bag was inserted into the container and the pH of any rain collected was determined using a BDH indicator strip. The quantity of daily rainfall, weather conditions and other environmental factors were also recorded. The data from the Hertfordshire and Norwegian monitoring were exchanged between schools and collated nationally by WATCH and the Hatfield Polytechnic.

Analyses of the WATCH Acid Drops Project, and the Hertfordshire–Norwegian contribution have been recorded elsewhere (Hiscock, 1988; Shirley, 1988, 1989), but some of the overall UK results are shown in Figure 13.1. Of particular interest to the Hertfordshire students was the high acidity, around pH 3.0, of local rainfall in areas where rivers and streams are noted for the high pH of their waters. Students from both countries were made aware from the results of the correlations between incidence of highly acidic rainfall, wind direction, frequency of rainfall, and location and type of industrial activity.

**Investigations into terrestrial and aquatic systems**

The Acid Drops Project results led to further studies of atmospheric pollution and the possible effects of acidification on a number of

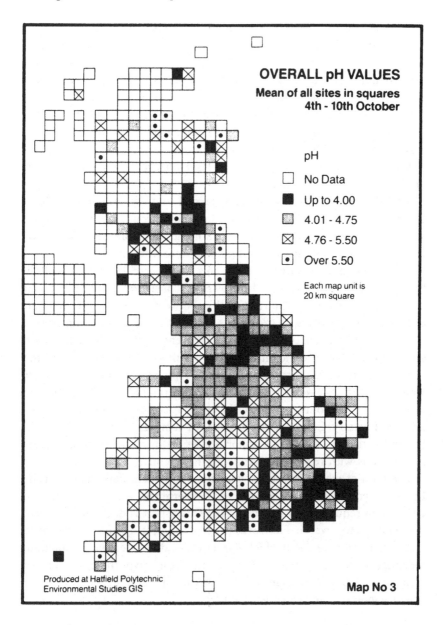

Figure 13.1. United Kingdom: Acid Drops 1986. Mean of all sites in squares, 4th–10th October.

organisms and communities. Attempts were also made to explain the mechanisms of the acid pollution cycle. In Norway, the students studied lake and river waters, measuring pH and assessing the biotic diversity in particular with regard to species whose acidic tolerance was known. Hertfordshire students carried out similar studies on ponds and streams.

Students in both countries researched earlier limnological records for the same water bodies for comparison. Not surprisingly, the Norwegian students found evidence, some of it anecdotal, of declining fish populations in lakes over a wide area whereas Hertfordshire students found no such decline in their areas.

Studies of woodland were also carried out by students in both countries. Investigations were made into coniferous and broad-leaved species and records made of crown thinning, discoloration and loss of leaves. Soil and laboratory investigations were also undertaken.

The results and experiences from this work were exchanged, chiefly by post, between linked schools. However, considerable care was taken in their interpretation. The relationship between acid precipitation and certain observed effects on organisms has not been clearly demonstrated. To show the complexity of the issues, students and their teachers were given further background information resulting from research by the Acid Rain Research project of the Norwegian Institute for Water Research, the Norwegian Monitoring Programme for Long Range Transported Air Pollutants, the UK Central Electricity Generating Board and the Forestry Commission. A series of seminars were also arranged in Hertfordshire to provide opportunities to discuss results and their implications with representatives from the scientific community.

The level at which individual students and classes participated in the broader ecological investigation and experiment was related to the curriculum requirements for their age. This aspect is dealt with in more detail later. However, in Hertfordshire, for example, further work ranged from examining the controlled effects of acidic water on germinating seedlings at the primary level, to constructing hypotheses about the route of acid precipitation in river systems and mineral precipitation from soils for post-16, advanced level students.

## The European Living Indicator Project

As part of the European Year of the Environment (1987–1988) the European Commission (EC) sponsored a major environmental education project involving six member states, namely, Belgium, the Federal

Republic of Germany, Italy, Luxembourg, the Netherlands and the United Kingdom.

The initiative for the project arose from the success of the 2nd EC Environmental Education Network, which finished in 1986 and involved only one school from each member state carrying out joint educational studies (Trant, 1987).

The aim of the project for the European Year of the Environment was to promote:

The environmental consciousness of the European youth that in a few year's time will contribute to the shape of Europe . . . to help young people to know their own environment through observation and investigation to better understand ecological relationships.
*(Hoffman, pers. commun.).*

It was based on the subject of 'Living beings as indicators of environmental quality in Europe'.

Participating schools were asked to investigate and make an ecological assessment of a local site, and to suggest how it might be improved by management, including habitat creation. This account deals solely with the contributions made by the United Kingdom, for which the author was the member state co-ordinator.

Schools in the United Kingdom were invited to select a suitable place for study. It could be the school's grounds or a nearby site with grassland, hedgerow, woodland or other features.

The aim for each school was the preparation of a report of the site containing the following sections:

1. a description of the site's physical features;
2. an account of the animals and plants found there;
3. suggestions for site improvement;
4. recommendations and plans for implementing these improvements.

Over 40 schools from England, Wales, Scotland and Northern Ireland entered the project. Most were primary but secondary and special schools (for students with learning difficulties) were also represented.

Most schools chose to study areas of between three to five hectares. Many of them were in school grounds, which they wished to develop as small nature reserves, others were areas adjoining the school, often described as 'eyesores'.

Guidelines were given to each school suggesting ways in which they could investigate and approach the ecological component inherent in considering areas 3 and 4 listed above. For many of the younger children these studies were observational and resulted in drawings and short written descriptions. Some attempts were also made at quantitative and qualitative assessment by primary schools of flowering plants, birds and invertebrate groups. Mapping of the sites and of vegetation was a common feature of all groups. The range of activities undertaken were:

Vegetation mapping
Pollination
Germination of angiosperm seeds
Mammal studies
Invertebrate studies:
  Mollusc diversity
  Insect surveys
Atmospheric pollution studies
Weather recording
Species distribution
Fruit and seed dispersal
Plant colonisation
Bird feeding behaviour
Freshwater studies:
  Pollution monitoring
  Pond and stream surveys
Soil analysis

Presentation of data in the reports varied from single drawings and brief descriptions from the youngest students to more sophisticated tabulations, diagrams and computer print-outs. In dealing with section 3 of the project, site improvement, all of the schools except two were concerned with habitat creation. In most cases this brought the students into contact with ecologists and nature conservationists from outside organisations, such as the County Wildlife Trusts and the British Trust for Conservation Volunteers. Several schools submitted detailed plans of their proposals, which frequently included the planting of native trees and shrubs, the creation of species-rich grassland and making ponds.

Although the project did not require the recommendations for improvement to be implemented, eight schools had already begun work on their site by the end of the report period.

A national report on the UK projects was submitted to the European Project Director at the end of 1988. The European Final Report, incorporating the results from all participating member states, was to have been produced and circulated in 1989 to all the schools that took part; however, problems in project completion prevented its appearance.

### Relevance to the curriculum

Both the Hertfordshire–Norwegian Acid Rain project and the European Living Indicators project took place before the introduction of the new National Curriculum in England and Wales. At that time the projects were seen as fulfilling part of school syllabuses concerned with science, geography and environmental studies. For the secondary school students the ecological component was used for public examinations in science and environmental studies for the General Certificate of Secondary Education taken at age 16, and for Biology Advanced level, taken at 18.

The National Curriculum Council for England and Wales (1990b) analysed the Hertfordshire–Norwegian links project. It concludes that the studies and general work of this kind contributes to several subject areas of the National Curriculum. Figure 13.2 shows the curriculum relevance of the project for pupils aged 11–14.

Ecology, as part of the new National Curriculum, is not a separate subject. It is a vital component of science and geography, and of the cross-curricular theme environmental education. However, it should be noted that the work described above meets the statement of the UK Department of Education and Science's policy statement (1985) that science should:

encourage curiosity and healthy scepticism, respect for the environment . . . and an insight into man's place in the world which will complement the contribution of other elements in the school curriculum

and of the Report by the Science Working Group for the National Curriculum (1988) that:

Science education should be concerned with the development of knowledge, with understanding, scientific skills and attitudes

### Discussion and conclusions

International projects such as the Hertfordshire–Norwegian link enable students to enhance their scientific development by contact with their collaborators in the partner-country. For the students, and teachers, this is a more meaningful method of gaining information and experience

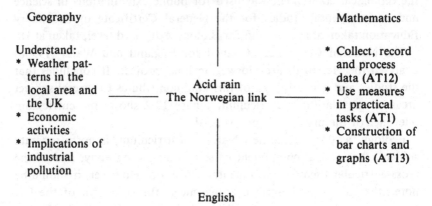

Science

* Carry out scientific
  experiments (AT1)
* Describe sources and
  implications of pollution
  (AT5)
* Understand relationships
  within the environment
  (AT2)
* Measure
  temperatures, rain-fall,
  wind direction (AT9)

Geography                                                          Mathematics

Understand:                                                        * Collect, record
* Weather pat-                                                       and process
  terns in the                                                       data (AT12)
  local area and              Acid rain                            * Use measures
  the UK             ——— The Norwegian link ———                      in practical
* Economic                                                           tasks (AT1)
  activities                                                        * Construction of
* Implications of                                                    bar charts and
  industrial                                                         graphs (AT13)
  pollution

English

* Write accounts of scientific investigations (AT3)
* Write letters (AT3)
* Select relevant information from various
  sources (AT2)
* Discussion/debate (AT2)

Cross-curricular areas

* Economic and industrial understanding
* Information technology
* Health education

Figure 13.2. The Hertfordshire–Norwegian link. Key stage 3 (11–14) analysis.
After National Curriculum Council (1990b). AT = Attainment Target.

than just using conventional teaching material in the isolation of their own schools. During the period of the two projects further international links have been established between schools in the UK and those in other parts of Europe through the School Links International (Beddis and Mares, 1988) programme aimed at students of primary age. Their results have led to similar conclusions. The use of electronic mailing and facsimile machines has also enhanced the communications possibilities between schools in different countries providing an extra dimension of immediacy, which is of considerable educational value.

If international projects between schools are to be of greatest benefit to students it is essential that their administration is efficient. It is clear, for example, that the delay in the completion of the Living Indicator Project has been to the detriment of the project's overall success.

Experience from the Hertfordshire–Norwegian link demonstrates that a sound infrastructure of co-ordinators and communication networks is vital. Given that, there is no reason why successful collaborative international ecological studies should not be a regular feature of the curricula in many more countries.

## References

Beddis R. and Mares C. (1988). *School Links International: A New Approach to Primary School Linking Around the World.* Avon County Council/Tidy Britain Group, Wigan.

Department of Education and Science (1985). *Science 5–16, A Statement of Policy.* London: HMSO.

Greig S., Pike G. and Selby D. (1987). *Earthrights: Education as if the Planet Really Mattered.* World Wide Fund for Nature/Kogan Page.

Greig S., Pike G. and Selby D. (1989). *Greenprints for Changing Schools.* World Wide Fund for Nature/Kogan Page.

Hiscock S. (1988). *WATCH Acid Drops.* Lincoln: WATCH Trust for Environmental Education.

National Curriculum Council (1988). *Interim Report by the Science Working Group for the National Curriculum.* London: HMSO.

National Curriculum Council (1990a). *Geography Working Group, Final Report, Geography for ages 5–16.* Department of Education and Science and the Welsh Office. London: HMSO.

National Curriculum Council (1990b). *Curriculum Guidance No. 8: Environmental Education.* London: HMSO.

Shirley D. (1988). *Acid Rain: The Norwegian Link. Environment 1988.* Hertford, UK: Hertfordshire County Council.

Shirley D. (1989). *The Hertfordshire–Norwegian Project Annual Review of Environmental Education No.2 (40–42).* Reading, UK: Council for Environmental Education.

Trant A. (1987). *The Outcomes and Possibilities of the Network*. Milieu
Bulletin of the European Community Environmental Education Network.
Special Edition for the European Year of the Environment. Trinity
College, Dublin: CDVEC Curriculum Development Unit.

# 14

# Creating a climate for conservation in West Africa

KEN SMITH

*International Officer-Africa, Royal Society for the Protection of Birds, The Lodge, SANDY, Bedfordshire, UK*

## Introduction

As Europe's largest non-governmental (NGO) wildlife conservation organisation, the RSPB is committed by its charter to the conservation of wild birds and their habitats.

The RSPB came into being in 1889 as a result of the attempts of a group of ladies to bring to an end the international trade in the feathers of wild birds. The campaign they launched aimed to make the public aware of the cruelty of this extensive trade and sought action on the part of those people who made up the RSPB's early membership. Each member signed a pledge not to wear feathers, a highly prized fashion accessory at the time. Lobbying was seen as an important part of the process which, some 30 years on, saw the plumage trade banned by Act of Parliament.

With the passing of the Plumage Act the RSPB's attention was drawn to other major issues concerning the protection of birds in the United Kingdom. Those early principles of awareness coupled with action have remained the corner stone of the RSPB's education programmes.

## The RSPB's West African projects

The RSPB is currently involved in three projects in West Africa:

1. The Save the Seashore Birds Project in Ghana (SSBPG)
2. The Hadejia-Nguru Wetlands Conservation Project in Nigeria (HNWCP)
3. The Gola Rain Forest Conservation Programme in Sierra Leone (GRFCP)

These have been operating since 1985, 1987, and 1990 respectively. The aim of these projects is primarily the conservation of wild birds. However, this has led to involvement in a large and complex range of human issues that demonstrate that the needs of wild birds cannot be addressed in isolation. The birds' habitats are almost always shared with humans whose needs and aspirations must be taken into account if the futures of both are to be secured.

The RSPB is involved in laying the foundations upon which ecological education can be built, creating a climate of public opinion that will allow conservation consciousness to be developed and harnessed in the future. In order to be of value ecological education has to address the real issues of development in Africa. It has to help people fit their own experience into the wider ecological picture in order that they can understand how their actions affect the local and wider environment.

People must be inspired to work for a future which is both sustainable and maintains biological diversity. The phrase 'think global, act local' is a cornerstone of the RSPB's approach to conservation education both at home and abroad.

It is a common misconception outside Africa that local people do not understand the problems that they face. This is far from the case as local people are usually well aware of the limitations of a particular area and the way in which it has changed over the years. Furthermore, they have frequently developed strategies that in the past alleviated some of the problems they had to deal with.

Recent developments such as increased population coupled with the breakdown in traditional systems that conserved limited resources have exacerbated already difficult situations. The 'technological fix' offered by the developed world has proved largely illusory. Africa has numerous failed multi-million pound projects illustrating that western development agencies do not necessarily have all the appropriate answers.

The solution to these problems does not always lie in transposing western conservation thinking to the developing world. Some of these concepts are completely alien, such that messages easily understood in the west are incomprehensible in Africa. For example, 'Save the Gola Forest' is a slogan readily understood in the developed world, but to the Mende people who live in and around the forest it is meaningless. The forest has always been there, it is their 'patron', it supplies almost all their daily needs and has always done so. This idea of 'patronage' is so strongly developed that it is difficult for these people to come to terms with the concept of protecting their 'patron'.

This is not to say that indigenous peoples do not understand the complexities of conservation, but that they often cannot relate to the terms in which they are couched.

Ecological education is a case in point. Other authors in this volume have alluded that ecology is a broad based subject that has been usurped by the biological sciences. This is true in the developed world. However, ecology embraces a wide range of disciplines, and these must be considered when designing programmes to develop public awareness to create a climate of opinion that is ready to accept conservation action. This is true at the local, national, and international level.

This essentially broad base of ecology was illustrated at the end of a training course for game protection guards conducted by RSPB staff as part of the Hadejia-Nguru Wetlands Conservation Project in the Sahel zone of Northern Nigeria in 1989. Participants on the course were asked why they considered it important to protect these wetlands. The subsequent discussion ranged over the wide range of topics listed below, considering the wetlands as:

1. a source of ground water recharge;
2. a feeding ground for migrant and resident birds;
3. sustaining economic activity such as farming, fishing, fuelwood collection, crafts and tourism;
4. a barrier to the desert;
5. a natural laboratory;
6. helping to maintain the area's ecological balance;
7. promoting international cooperation;
8. providing drinking water;
9. a transport system;
10. maintaining soil fertility;
11. providing aesthetic inspiration to artists and craftsman;
12. promoting self employment and therefore self-esteem;
13. part of the national heritage;
14. a source of medicinal plants;
15. contributing to climatic amelioration.

The above list gives an insight into the factors that have a bearing on the conservation of (in this case) birds and their habitat. It is against this broad ecological background that the RSPB has tried to develop its environmental education programmes.

## Ghana

The Save the Seashore Birds Project (SSBPG) in Ghana is the longest running RSPB sponsored project in West Africa. It was established to conserve Britain's rarest seabird, the roseate tern (*Sterna dougalii*). During the early 1980s the tern's population in Western Europe was declining despite the fact that its breeding grounds (particularly in the British Isles) were largely protected. 'Ringing' recoveries indicated that the source of the problem might be in West Africa, and Ghana in particular. It was thought that the trapping of terns by children could be a critical factor in the species' decline.

Representations were made to the government of Ghana who were concerned about the reduction in tern numbers, but had no resources to devote to a low priority area such as bird conservation. Working with concerned individuals in Ghana, the RSPB established the Save the Seashore Birds Project (SSBPG) as, at that time, there were no Ghanaian non-governmental organisations (NGOs) through which to work. A project agreement was signed between the RSPB and the Government of Ghana. This process of signing a project agreement with the government of the country in which a project operates has been followed by all subsequent projects. The RSPB then works with a conservation NGO in that country.

The initial aims of the SSBPG were threefold:

1. to monitor the numbers of terns, particularly roseate terns, wintering along the coast of Ghana;
2. to identify the most important sites used by the birds in order to be able to protect them at a later date;
3. to assess the impact of tern trapping and initiate a conservation education programme to address the problem.

The first of these aims is ongoing, and monthly counts are made throughout the year on the most important sites. Full coastal surveys are being undertaken three times a year. The most important sites have now been identified: six areas are used by up to 80% of the wintering tern population.

The third aim has been the greatest success but at the same time a source of frustration. A failure to collect baseline data on tern trapping has resulted in an inability to assess the true effectiveness of the education programme objectively.

The project has undoubtedly been successful in raising public awareness of conservation issues in Ghana and has helped create a climate of opinion in which conservation action is being undertaken at all levels of society. However, it is not possible to quantify the success of the programme with regard to its impact on tern trapping due to unreliable data in the initial stages of the project.

A new survey of trapping is now underway, which will be of value both in assessing the current extent of the problem and in targeting future areas for the education programme to work in. Nevertheless, as is so often the case, an opportunity to measure the effectiveness of an environmental education programme has sadly been lost.

The education programme has two objectives:

1. to inform and educate the local population about the importance of the terns, including roseates, and obtain a commitment to cease trapping them;
2. to set up conservation clubs in schools of the coastal region, operated by local volunteers.

The clubs were modelled on the Young Ornithologists Club (YOC), which is the junior section of the RSPB. Although there were initial difficulties there are now 150 clubs, with 7000 members in total, successfully established. The 'Wildlife Clubs of Ghana' as they have become known have been very significant in promoting conservation awareness both locally and nationally.

The clubs aim to be self-sufficient, members pay a fee to join and receive a membership card and certificate. They develop their own constitutions and raise funds both for their own running costs and to donate to centrally run projects. A club magazine (*NKO*), loosely based on the YOC magazine *Bird Life*, is published to keep members in touch with issues and involve them in competitions and surveys. Club members are encouraged to take an active part in the production of the magazine by writing articles and letters for publication.

The 'Wildlife Clubs of Ghana' have also developed a number of high profile events to raise the general awareness of conservation by catching both public and media attention. The National Wildlife Week held in November 1991 is an example of an event that has helped to raise public consciousness of the need for conservation at both local and national level.

The major events of the National Wildlife Week, run in conjunction with the Government's Department of Game and Wildlife, included the launch by the Secretary for Lands and Natural Resources, of Ghana's first sponsored birdwatch, and the relaunch of the Ghana Wildlife

Society (a previously defunct NGO). A fundraising dinner was hosted by the First Lady of Ghana and there was a march through Accra by 1000 members of the Wildlife Clubs of Ghana. These events were followed by a funfair and presentations by government ministers. Widespread press, radio, and television coverage ensured a wide audience for the events.

The launch by the Secretary for Lands and Natural Resources and the involvement of other high ranking government officials brought the project and the government together, demonstrating that governments and NGOs can work together to achieve benefits for both wildlife and the community. The sponsored birdwatch showed that children in developing countries can take an active role not only in the promotion of conservation issues but also in practical fund raising to support conservation action.

This raising of conservation awareness and the careful development of the SSBPG has led to interest from the Global Environment Facility (GEF) of the World Bank. A proposal by the project, supported by the Ghana Government, to create a series of refuges along the Ghana coast was viewed favourably by the (GEF). SSBPG has carried out much of the initial groundwork on which the programme will be based. The sites were already documented and three have management plans written as part of a previous RSPB sponsored training course. Considerable data were available and an education/public awareness programme was in place. The high degree of professionalism shown by the project's Ghanaian staff assisted the creation of a concise project plan for the World Bank.

The World Bank has sought a guarantee from the Ghana Government that it will, as a signatory of the Ramsar Convention, designate five sites as Ramsar sites. These will have their bird populations protected while at the same time allowing sustainable human use. It is likely that the government will accept this condition, resulting in the sites becoming permanently protected areas.

The project's future plans include the creation of a conservation education centre, which will incorporate the offices of the Ghana Wildlife Society and the Wildlife Clubs of Ghana as well as lecture facilities and nature trails.

## Nigeria

The RSPB's conservation and education work in Nigeria differs from that in Ghana and is more complex. The Hadejia-Nguru Wetlands Conservation Project (HNWCP) was established to conserve the wetlands of the Hadejia-Jama'are floodplain of Northern Nigeria. These areas are close to the southern fringe of the Sahara Desert and are the winter home of thousands of European migrant birds as well as the breeding grounds for large numbers of migrants from within Africa.

As in Ghana, a project agreement formed the basis of the project. The signatories were the RSPB, the International Council for Bird Preservation (ICBP), and the Federal Government of Nigeria. The project works through the Nigerian Conservation Foundation (NCF), the country's national wildlife conservation NGO, and the World Conservation Union (IUCN).

Mapping and bird monitoring was initially undertaken, followed by the appointment of an education officer. However, it was some time before the education officer was able to organise an effective public awareness programme and this delay resulted in major misconceptions about the project among the local population, which caused a number of problems. This illustrates the importance of having an education officer in place at the start of a project.

Like Ghana, the project is based in a densely inhabited area (up to one million people), with the attendant pressure on land. The ornithological objective of the project alone was not a powerful enough argument to gain local support. The pressure on land in the region was high and increasing and any plan to conserve the wetlands had to take account of human aspirations and development needs.

The threats to the wetlands are complex. They include the Sahel droughts of the last two decades and human-induced problems such as major dam schemes (either in place or planned) for the upper reaches of the rivers that supply water to the wetlands. These problems are compounded by environmentally degrading actions including over-fishing, over-grazing, over-exploitation of ground and river water, changes in land tenure, and a breakdown in traditional systems of land use. The latter helped to maintain the delicate balance between the diverse interests of the area's population.

Against this background the project's early attempts at conservation education had little effect. Political demands exacerbated the problem. Early in 1989 the Governor of Borno State decreed that all schools should have conservation clubs and that one member of staff should be responsible for its organisation. This seemingly laudable action resulted in teachers without the necessary experience or training running the clubs. This forced the project to divert its efforts into training club leaders, even though the appropriateness of the clubs in the Nigerian context had not been evaluated.

A training course for leaders was sponsored by the African Wildlife Foundation (AWF) and relied heavily on their experience in East Africa. Unfortunately, this input proved unsustainable. This course remains the only in-service training for teachers run by the project to date. Much was learnt from this experience, especially the importance of targeting in a country as large as Nigeria. Additionally, it highlighted the importance of gathering baseline data on which to construct an education/awareness programme and against which to evaluate its results. The Ghanaian experience has highlighted the importance of an evolutionary process in education programmes.

For a considerable time the awareness programme was based on a set of slides produced by the project's education officer at the International Centre for Conservation Education (ICCE) in the UK. This was only partially successful as the slide set proved difficult to update and adapt to the needs of different communities within the wetlands.

However, two notable successes were achieved by the project. The first of these was in training game guards and the second was the development of strong links with the Ministry of Education in what was then Kano State. The latter resulted from the enthusiasm of the Kano observer on the original AWF course, who lobbied his State ministry to set up a conservation education unit, which it subsequently did. After training in the UK (sponsored by AWF) he returned to Kano to implement conservation clubs successfully in the State's secondary schools.

The game guards courses are innovative and effective. The objectives are threefold: to enable participants to identify birds; to enable them to monitor changes in bird numbers; and to train them to act as 'extension' workers when they returned to their own communities. Initially this training was undertaken by a member of the RSPB's staff. A modular course was devised which is now taught by the project's local staff. These game guards courses are now organised annually by the project's Wildlife Survey Officer and approximately 75 guards have been trained and equipped with field guides, notebooks, and binoculars. Some are now involved in the annual waterfowl count run by the International Waterfowl and Wetlands Research Bureau (IWWRB).

The future of the project is now being planned. The hydrology and economics of the area have been researched to provide a basis to argue in favour of the retention of the annual flood upon which both the human and bird populations depend. A new education programme is in preparation and a new Education Officer has been appointed.

## Sierra Leone

The Sierra Leone project in the Gola forest has been severely curtailed by the war in the neighbouring Liberia which has spilled over the border. In the intervening period an informal education programme is being developed. It will take as its model the Wildlife Clubs of Ghana, and the SSBPG's education officer has travelled to Sierra Leone to provide support and advice. The Nature Clubs of Sierra Leone was launched in autumn 1992.

## Training

The RSPB has provided training opportunities for project staff both in the UK and onsite. The ICCE's introductory course in conservation education provides a good grounding in community education and public awareness techniques. It has developed over the years to address the requirements and backgrounds of students and their future roles. In addition the RSPB customarily arranges a short placement with a UK conservation education organisation, providing an opportunity for students to practise their new skills before they return home.

Individual study tours to UK institutions are organised by the RSPB's International Training Officer. Occasionally, specialist courses, such as the 12 week course at Jordanhill College of Education in Glasgow, are arranged. This course has been developed by the college in conjunction with the World-Wide Fund for Nature (WWF) for education officers who are in a position to influence the school curriculum in their home countries.

Increasingly, courses are being undertaken in the home country of the project. This is seen as both more effective and economical. The RSPB's policy is to promote visits by project staff to other projects to enable them to share and benefit from each others' experiences.

## Resources

There is a shortage of suitable educational resources for all the African projects, particularly of written and visual materials. Lack of staff time in the early stages of projects is usually the main constraint.

In Ghana a start has been made with the publication of *NKO*, the magazine of the Wildlife Clubs of Ghana. This is distributed to secondary schools, government offices, embassies, and major companies. The magazine is currently funded by the RSPB but is hoped it will secure its own source of funding in due course.

## Conclusion

The RSPB has been undertaking conservation projects in West Africa since 1985. The educational elements of these projects have been variable in terms of their success. Many of the RSPB's bird conservation priorities lie outside the UK and it is likely that projects will continue to be developed to address those priorities. It is therefore vital that conservation education forms a basic part of these projects and that the RSPB builds on its experience in the field to promote awareness of the issues affecting the conservation of birds coupled with action that leads to the wise use of the environment by its human population.

# Index

Acid Drops Project [UK], 170
Acid Rain Project [Norway], 172
African Wildlife Foundation, 186
AIDS, 25, 31
Aids to Identification in Difficult Groups of Animals and Plants Project (AIDGAP) [UK], 78
Alliance for Environmental Education [US], 49
American Association for the Advancement of Science (AAAS), 48
American Forest Foundation, 48
Association of School Nautral History Societies [UK], 83

Basic Education in Integrated Rural Development Project [Uganda], 27
Biological Science Curriculum Study [US], 54
British Association for the Advancement of Science (BAAS), 72
British Drug Houses (BDH), 170
British Ecological Society (BES), xii, xiii, xv, 11, 44, 60, 79
British Petroleum (BP), 170
British Trust for Conservation Volunteers (BTCV), 174
Brundtland, Gro Harlem, xii
Brundtland Commission, 10
Brundtland Report, 20
Buchanan-White, Francis, 61

Carry American Relief Everywhere (CARE), 30
Central Electricity Generating Board (CEGB) [UK], 169, 172
Centre for Environmental Research and Innovation, 86
CFC's, 39
Chiba Prefecture [Japan], 163

Children Alliance for the Protection of the Environment (CAPE) [Indonesia], 159
Conservation Association of Miura Peninsular [Japan], 163
Cornell University Environmental Sciences Interns Programme [US], 52
Council for Environmental Education (CEE) [UK], 85
Council for the Preservation of Rural England (CPRE) [UK], 83
Countryside Commission [UK], 122
cross curricular theme, 15, 18

Danish Agency for International Development (DANIDA), 30
Davenport, Iowa [US], 48
Denver Audobon Society Urban Ecology Project [US], 52
Department of Education and Science (DES) [UK], 122
Department of Game and Wildlife [Ghana], 183
diarrhoea, 31
Dobson, Sissel, 169
Duke of Edinburgh Award Scheme [UK], 84

Eco-Inquiry Curriculum [US], 54
Ecological Society of America (ESA), 45, 49, 50
Ecology, Congress of, xi
Ecology, Institute of, ix
Edo River [Japan], 164
Edogawa Ward [Tokyo, Japan], 164
Education Policy Review Commission [Uganda], 28
Education Reform Act, 1988 (ERA) [UK], 10, 11, 16
English Nature [UK], 139
Environmental Impact Analysis (EIA), 145, 150